Municipal Control of Cable Communications

Robert E. Jacobson

The Praeger Special Studies program—utilizing the most modern and efficient book production techniques and a selective worldwide distribution network—makes available to the academic, government, and business communities significant, timely research in U.S. and international economic, social, and political development.

Municipal Control of Cable Communications

PRAEGER SPECIAL STUDIES IN U.S. ECONOMIC, SOCIAL, AND POLITICAL ISSUES

Praeger Publishers New York London

Library of Congress Cataloging in Publication Data

Jacobson, Robert E 1948-
Municipal control of cable communications.

(Praeger special studies in U.S. economic, social, and political issues)
Bibliography: p.
1. Community antenna television. 2. Telecommunication. 3. Municipal government. I. Title.
HE8700.7.C6J3 1977 384.55'47 77-7815
ISBN 0-03-021831-4

PRAEGER SPECIAL STUDIES
200 Park Avenue, New York, N.Y., 10017, U.S.A.

Published in the United States of America in 1977
by Praeger Publishers,
A Division of Holt, Rinehart and Winston, CBS, Inc.

789 038 987654321

Printed in the United States of America

PREFACE

Despite claims made on its behalf, the cable television medium has not provided the cities with a level of service expected by communication policymakers. Several causes have been given for this failure on the part of the new medium. At the same time, the cities, which are the prime clients of the cable medium, are experiencing severe difficulties in their struggle to remain healthy; the urban working classes and poor, especially, have had to bear the brunt of municipal inefficiency and generally oppressive conditions. It has been claimed that advanced cable communications systems could avert further urban collapse by restoring democratic and efficient management to the cities, and by stimulating economic recovery. Should the cities take the initiative in developing municipal cable communications systems? This is the central problem of this book, an examination of possible planning options available to American cities today. The historical development of the cable medium in an urban context, the possible role of telecommunications as a public good, the failure of the private entrepreneurs to fill this role, and means for the cities to have cable fill this role are investigated.

The study concludes that cities should undertake the development of advanced telecommunications systems employing cable technology and that such development *must* incorporate active and intense public involvement at each stage of planning and implementation. Further, citizen involvement is seen as a vital first step in the reorganization of urban society, so that the full potential of the communications infrastructure can be exploited by the people of the city in an ongoing and progressive social transformation.

ACKNOWLEDGMENTS

Books such as this are not the work of any single person, though the author takes final responsibility for the conclusions of this study (and any errors in its midst). Cooperation from many parties made its completion possible.

Without the assistance of many city officials throughout the United States, and most notably San Bruno's City Manager, Gerry Minford, and Director of Public Works, Leo Van Dusen, there were be little substance to many of the author's claims for public cable systems. Together with John Bateman of the Public Cable Television Authority, they have shown that public telecommunications can function and function *well.*

Harold Horn of the Cable Television Information Center greatly aided while aiding the author in his researchings.

My thesis committee deserves special thanks. Each contributed more than the mandatory reading and critiquing of this study. Geoffrey Cowan, an early advocate of sovereignty for the public in communications and former director of the Communications Law Program at UCLA, pointed the author in the right direction when his graduate program was faltering. Mark McCarty, truly "a man of the people," initiated a community-building project in Topanga Canyon that demonstrated to the author the social utility of the video medium when properly used. Thesis committee chairman Walter Kingson was a mature and thoughtful critic (in the best sense of the term), who forced this study to take coherent form, despite preceding months of chaos.

One other friend at UCLA, Professor Mitsuru Kataoka, first exposed the author to the potential of the video medium as an end in itself—and provided a properly Zen philosophy that allowed the author the personal luxury of taking the time to do this task right.

Additionally, two fellow students at the Annenberg School of Communications at the University of Southern California contributed in the acute, final months of this work. Fred Horstmann, now at Simon Fraser University, helped the author to formalize his earlier "intuitions" into more accurate critical analyses, while Dan Wedemeyer edited an early draft of this book and made valuable and incisive comments. Carole Szymecko labored long nights to complete typing the final manuscript.

Finally, great thanks must go to my parents—all four of them. Their support and encouragement never faltered over two long and arduous years. My greatest appreciation is reserved for my wife, Denise. Her pride in this accomplishment overwhelms my own—and that is the best reward I could hope for.

CONTENTS

Municipal Control of Cable Communications

CHAPTER

1 THE PROBLEM AND RESEARCH DESIGN

THE PROBLEM

The growth of cable television in the United States over the last 20 years has been prodigious: from a handful of systems in the 1950s, by 1975, the industry had grown to encompass over 3,000 operating systems.[1] But many of the claims made for the new cable medium—above all, that it would allow the development of extremely efficient and democratic "wired cities"[2]—have not been fulfilled by the cable industry. Instead, the system has come to resemble an entertainment adjunct to conventional television and motion picture media.[3]

One comprehensive and authoritative source has laid the blame for this sub-optimal use of the cable medium at the door of the Federal Communications Commission (FCC).[4] Others have attributed the triviality of current service offerings of the industry to gross venality on the part of cable operators, while the operators themselves have been quick to point the finger at "obstructionists" among the established television and telephone industries.[5] Finally, a well-known communication-policy researcher has imputed the slow development of interactive and other innovative capabilities to a general public lassitude with regard to communication issues, and especially those related to cable communications.[6]

The present plight of the medium is probably a result of the interaction of all of these causes, plus the inability of local regulators—cities and states—to come to grips with the complex issues surrounding cable development.[7] For example, should cable be seen as a tremendous boon to urban efficiency? Or should it be viewed as merely a transient intruder in the communication technology field, which has as its sole appropriate function the supply of premium TV services to affluent suburbs?

Meanwhile, the cities themselves seem to be involved in a general and continuous crisis of managerial failure, loss of citizen morale (leading to an exodus of residents and industry), and local unrest.[8] This crisis has been connected with a variety of dynamic urban phenomena, including rampant technology, implemented without thought for the human consequences.[9]

Seymour Mandelbaum, however, has concluded that humane planning for urban development is hampered by the lack of adequate communication technology in the city; the growth of active communities within the city could be stimulated and supported by interactive cable communication systems connecting their constituents.[10] Mandelbaum has enunciated a policy that would have cities (possibly with federal help) advance the capital for the construction of these systems and, in effect, run them as public utilities under public ownership.

Should the cities take the initiative in developing municipal cable communications systems? This is the central problem of this book.

In answering this question, the historical situation of cable television must be analyzed to discover why conventional cable television has not fulfilled promises made on its behalf. The concept of public ownership of a medium of communication is fraught with controversy in the American context; its justification demands a demonstration that traditional private entrepreneurs cannot provide as high a level of service as a public enterprise in an urban environment. Beyond questions of ideology, what steps could the cities take to ensure the most efficient use of public resources in supporting a public cable communications enterprise? What alternative forms of planning and implementation are available to the cities, and what are their advantages and disadvantages?

Ultimately, the crisis of the cities—and, possibly, the success of interactive telecommunications systems—will require human solutions to human problems. The reorganization of urban society would be a necessary prerequisite to the success of technical solutions to the urban crisis. How can this reorganization take place, how can it complement the introduction of cable communications, and how can it work to guarantee the efficient and democratic operation of the cable system so as to prolong the efficacy of the reorganized social structure in return?

This study will attempt to answer these questions and, in so doing, set an agenda for city officials and urban citizens concerned with finding solutions to the urban crisis and gaining a better understanding of the role cable communications could play in implementing those solutions.

SIGNIFICANCE OF THE STUDY

Among the multitude of studies undertaken to better understand the peculiar dynamics of cable communications in the past 15 years, none has adopted a truly critical perspective as part of its analytical scheme. These studies reflect a too easy acceptance of the status quo as the probable future; their authors are, as Herbert Schiller has put it, "waiting for orders"—rehashing the past in order to support the oppressive plans of exploitive interests in the society with "scientific" facts.[11]

City officials and citizens working with innovative technologies need a less restricted vision of what the future might be like. The information they require to plan and develop cable must be just as valid, based on just as discernable empirical facts, as the more limited extrapolations of noncritical studies. However, when facts do not exist—or, rather, when quantifiable data are not readily available—a qualitative analysis based on an historical overview and an evaluation of present social and economic trends can reveal artificial constraints on planning that could and should be discarded.

Citizens organizing around cable issues might benefit from this discussion in another way: while most cities (and more each year) consider public ownership of the cable medium as one desirable option in planning for urban telecommunications, there is no source that adequately defines the issues sure to come up during deliberations. Neither the FCC nor state agencies have been prepared to offer counsel. The Cable Television Information Center and other third-sector groups have also been unwilling to advise cities considering public ownership to move in one direction or another.

Citizens need to know specific procedures for analyzing public versus private ownership in the local context; such advice can be found in the excellent booklet (no longer in print) published by the San Gabriel Valley Public Cable Council, *Municipal Decision-Making Factors Relative to Cable Television Ownership.*[12] They must also be concerned with the broader issues, which seldom get discussed in the public forum—the fantasy world of popular ideology and its effect on deliberations, the failure of the private cable industry to supply even conventional service in a reliable manner, and the needs of the cities for more than conventional service. This study attempts to define *these* issues, following an examination of their historical origins.

In the battle against oppression by owners of the media in America, the presentation of alternatives to the status quo can be powerful weapons in liberating men's and women's minds from debilitating acceptance of societal cliches. Such propositions can easily degenerate into utopian vistas, whose inability to be realized is even more destructive of collective action than were no proposition to be advanced at all. It is the author's contention that the concept of municipal cable communications as a vital part of the urban environment is not mere conjecture but a completely possible development in urban America—and one that will come sooner, in a progressive and democratic form, with greater public awareness of its possibility.

REVIEW OF THE LITERATURE

Reports and publications dealing with cable television began to appear in the mid-1960s, beginning with the FCC's decision to impose a moratorium on cable television growth in 1966. By 1970, the frequency of these items was beginning to peak, while the appearance of the industry magazines, *CATV* and *TV Communications*, and the alternative journal, *Radical Software*, indicated the growth of a large and marketable audience. However, for all the variety and quantity of these reports–including several dissertations, most assessing local government regulation or the use of access channels–few dealt with the issue of public ownership of the media.

The richest sources of information were the antagonists in the struggle for dominance in the area of local communications. The National Cable Television Association (NCTA) provided a large packet of articles and newspaper clippings defending their viewpoints on the topic. Equally prodigious mailings were forthcoming from PubliCable, Inc., the cities of San Bruno and Palo Alto, and the Public Cable Television Authority of Orange County–all in California. Also forwarding extremely informative materials were the cities of Detroit, Jacksonville, Florida, and Milwaukee. The province of Saskatchewan, Canada, and the government of Finland were the foreign sources of information.

It would be a mistake to overlook the circulars and handouts graciously offered by many of the smaller cities with municipal cable television systems already in operation. Only their very modest circumstances prevented them from offering further aid. However, the general brevity of these materials precluded their use in the completion of this study.

RESEARCH DESIGN

The method used for this study was descriptive-critical. The author's own involvement in the "video movement" of the early 1970s allowed for an intimate association with the cable television movement and its public interest critics for the past five years. This association made it possible to receive many research reports and other materials otherwise undocumented in standard research guides.

The descriptive portion of this study was compiled from numerous published and unpublished materials. It attempts to describe the historical development of cable television, indicating probable trends for the future development of the medium under the status quo. It also seeks to illustrate the immediate situation of the cities and makes a hypothetical connection between the suboptimal development of cable communications and the oppressive nature of city life.

The critical portion of this study derives from a dialectical method gaining favor among communication researchers in Canada and Scandinavia and, to a lesser degree, the United States.[13] It demands the recognition of systemic structures and processes comprising the phenomenon under investigation and the discovery of contradictions in its composition.[14] In the present study, the cable industry and its urban clients are seen as the phenomena requiring examination in a system context.

Initially, it was thought a good source of data would be available from cities already having publicly owned cable television service. A survey was constructed and distributed among the approximately 20 cities having such service. Response, however, was poor—only four surveys were returned, and these were, for the most part, incomplete. Limiting subsequent queries to the larger cities of San Bruno, Palo Alto, and Frankfort, Kentucky, it was possible to obtain virtually all the available information relating to present-day operations of cable television in a public ownership context.

The author also visited the offices of the Public Cable Television Authority in Orange County, California, and attended meetings of that body in 1975-76. Considerable insight into the legislative and political processes triggered by public ownership plans was obtained through such liaisons.

The major difficulty encountered in the preparation of this study was, of course, the absence of an actual large-scale publicly owned system available for rigorous examination. Even the critical method is inadequate to the task of foreseeing the *actual* future of a phenomenon; the richness of human experience defeats attempts to isolate particular variables from which concrete predictions of the future can be made. The more complex the system under investigation, the less reliable such prediction becomes. Therefore, this study does not presume to instruct city officials and citizens how public cable communications *must* come about but only to suggest issues that will present themselves in most urban environments and which must be dealt with according to the particular needs and resources of each locality.

NOTES

1. Herbert S. Dordick et al., *A Study of Local Government Uses of Cable Television and Integrated Communication Systems* (Los Angeles: Center for Communication Policy Research, University of Southern California, 1975).

2. Ralph Lee Smith, *The Wired Nation: Cable TV, the Electronic Highway* (New York: Harper & Row, 1972).

3. The Network Project, *Cable Television* (New York: The Network Project, Columbia University, 1973), notebook no. 5.

4. Don R. LeDuc, *Cable Television and the FCC* (Philadelphia: Temple University Press, 1973).

5. The Network Project, *Cable Television: End of a Dream* (New York: The Network Project, Columbia University, 1974), notebook no. 8.

6. Walter S. Baer, *Cable Television in the United States: Revolution or Evolution?* (Santa Monica, Calif.: Rand Corporation, 1974).

7. Dordick et al., *Local Government Uses of Cable Television*.

8. Duane Elgin et al., *City Size and the Quality of Life* (Washington, D.C.: National Science Foundation, 1974).

9. Serge Chermayeff and Alexander Tzonis, *Shape of Community* (Baltimore: Penguin, 1971).

10. Seymour J. Mandelbaum, *Community and Communications* (New York: W. W. Norton, 1972).

11. Herbert I. Schiller, "Waiting for Orders–Some Current Trends in Mass Communications Research in the United States," *Gazette* 20, no. 1 (1974): 11–21.

12. San Gabriel Valley Public Cable Council, Pasadena Urban Coalition, and Beverly Hills Bar Association Law Foundation, *Municipal Decision-Making Factors Relative to Cable Television Ownership* (Pasadena, Calif.: Pasadena Urban Coalition, 1973).

13. Kaarle Nordenstreng, "From Mass Media to Mass Consciousness: Current Thinking in Scandinavia," in Yrjo Littunen, Sirke Minkkinen, and Kaarle Nordenstreng, *Approaching Mass Media Research Through Communication Research* (Tampere, Finland: Institute of Journalism and Mass Communication, University of Tampere, 1974).

14. Anthony Wilden, *System and Structure* (New York: Tavistock, 1972).

CHAPTER

2

BACKGROUND TO THE OWNERSHIP CONTROVERSY

With regard to the immediate urban situation, one must qualify the boldest measures by a realization that they must overcome a massive inertia: indeed, they must go against the dominant forces in our civilization, even those forces in science and technology which, if we eventually emerge from this Age of Violence, may be at least put to more admirable human uses.

Lewis Mumford, *The Urban Prospect*

This is a free capitalistic society. I cannot see why you should have municipal ownership and operation of CATV systems any more than a municipality should run a newspaper, a motion picture theater, a restaurant or a dry cleaner.

Richard Loftus, in MITRE Corporation, *Symposium on Urban Cable Television*

Despite the long and successful history of municipally owned and operated enterprises—in fields as disparate as law enforcement and the supply of electrical power—proponents of "free enterprise" have initiated and sustained a concerted campaign to discredit the concept of publicly owned communications in the city. Private monopoly control of the means of communication on a national scale has long been the case in the United States, ever since David Sarnoff began assembling his networks under the protection of the Radio Act of 1927;[1] even earlier, one state after another had succumbed to domination by the Bell System, after first establishing "public utility commissions" to give the guise of protection of the public interest.[2] These conditions have sustained the public impression that no alternative forms of media management have ever, or can ever, exist.

But the campaign against public ownership of urban cable communications has taken on special meaning as more cities have seriously begun to examine public ownership–through the municipal entity–as one sure means of extending cable services to their citizens. The antipublic ownership forces have resuscitated arguments laid to rest in the era of the robber barons, when they were used to battle such evils as municipal street lighting.

These arguments have been advanced in defense of a private cable industry that has delivered on few, if any, of the promises it made, while speculating on apparently lucrative cable franchises in the 1960s and early 1970s. These "dangers" include the alleged inability of public agencies to successfully manage new technologies, the "luxury" nature of these technologies in capital-pressed cities, the political threat to freedom of speech "inevitable" in government control of any media, and the chance that government ownership might usurp the role of private enterprise in the provision of services–in short, the possibility that American capital might be prohibited entry into a new market laid open to exploitation by an FCC lax in its protection of public resources and an Office of Telecommunications Policy (OTP) intent on abetting such explotiation.[3]

Such arguments, for all their vehemence, ignore the historical situation of an American cable industry that is operating obsolescent, five- and ten-year-old systems, providing little more than the retransmission of conventional television signals and a few pay-movies and pay-sport programs: with over 3,000 communities wired for cable television, fewer than 50–or about 1.6 percent–are capable of providing two-way services (though none presently do so in any commercial quantity).[4] This situation has been frozen into law by regulation of the FCC, which has recently exempted pre-1972 systems–the vast bulk of operating franchises–from provision of two-way services.

The conflict between proponents and opponents of public enterprise is not an insignificant battle in the development of the infant cable communications medium. It is, Sackman has described, a major struggle over what may become the richest prize imaginable . . . the concentration of information power embodied in mass communication utilities," even though "the stake of the general public, which overshadows the economic struggle, is either generally neglected or only grudgingly acknowledged by the major contenders."[5] Sackman's primary concern is for the use of computer utilities, yet his comments apply equally to cable communications. The two are intimately intertwined: in the near future, cable will be the medium of choice for the transmission of computer information, while the demand for access to that information will encourage the extension of cable to many more subscribers.

John McHale has summarized the challenge facing our communications-dependent society: to clearly and comprehensively evaluate the consequences of public responsibility–or irresponsibility–in managing the development of the means of communication:

> What the new information environment actually does is to set the consideration of "issues" and approaches to problems in a quite

> different context. It is one in which we cannot tackle one issue, one question, or one problem at a time as isolated items . . . in which the perturbations of one component in one subsystem can grossly affect other components in ways that are quite different from the predictable behaviors of any isolated part of the system.[6]

There is a need to communicate this need for public responsibility to the urban public, to gain support for municipal entry into the field of urban telecommunications and to ensure the responsiveness of the communications infrastructure of the city to changing public needs. The maximization of participatory decision making and agenda setting in the urban context—made possible by an urban cable communications system—could vastly improve the city as a social entity—whose primary purpose is to be a good place to live for its inhabitants. The current crisis of the American city, as exemplified by New York City (which has more native resources than most cities, yet which hovers on the brink of complete disintegration), demands radical solutions to long-standing problems of social and economic inequities compounded by gross urban mismanagement.

A cable communications infrastructure alone will not suddenly transform the city into a better place to live. For citizens to effectively design and use the system and the information it can transmit will probably require extensive alterations in the social structure of the city. Such alterations have been, in the past, either uncontrolled (as with the inflow of substantial numbers of immigrants) or controlled by a small economic elite (as in the case of urban "redevelopment"—mandated plunder of land and homes serving a defenseless poor). Cable communications could make possible the self-determined alignment of urban residents into communities of common needs and interests—whether geographical or otherwise—but only with the involvement of citizens in every phase of development and management of the cable communications system could this capability be realized.

FCC Commissioner Benjamin Hooks, discussing the future of cable communications, predicted his own personal disappointment "if cable television—despite its historical braggadocio about new and different services . . . is to be no more than a passive conduit for distant television signals or a wired box office for pay-cable movies and sports."[7] Cable *can* be more—an integral part of the reconstruction of the American city. The time for action is now: a radical change in the structure (both physical and social) of the city and its mode of operations may be possible for only a short while longer. Publicly owned urban telecommunications could be the catalyst allowing the city to realize its full potential as a human—and humane—invention.

NOTES

1. Erik Barnouw, *A History of Broadcasting in the United States*, vol. 1 (New York: Oxford University Press, 1966).

2. Richard Gabel, "The Early Competitive Era in Telephone Communication, 1893–1920," *Law & Contemporary Problems* 34 (1969): 340–59.

3. Donald O. Williams and Spencer R. Kaitz, *Cities Should Not Own CATV Systems Because . . .* (National Cable Television Association).

4. Herbert S. Dordick and Carl Pilnick, *Integrated Urban Telecommunications for the Delivery of Municipal and Social Services in Los Angeles—Design for System Interconnection*. Los Angeles: Center for Communication Policy Research, University of Southern California, 1975), p. 1.

5. Harold Sackman, *Mass Information Utilities and Social Excellence* (Princeton, N.J.: Auerbach, 1971), p. 3.

6. John McHale, "The Changing Information Environment: A Selective Topography," *Ekistics* 211 (June 1973): 327.

7. Federal Communications Commission, Docket no. 20508, FCC 75-644 (released: June 27, 1975), concurring opinion, p. 6.

CHAPTER

3

CABLE AND THE CONTEMPORARY CITY

THE STATE OF THE CITY AS A STATE OF MIND

The environment in which one lives is a major determinant of one's ability to function and grow as a human being. The American urban environment of the 1970s has contributed significantly to the development of contemporary American life and thought.

What environment is the American city? According to a recent report issued by the Stanford Research Institute (SRI) on behalf of the National Science Foundation the large city is getting larger—but the increasing scale of living environments runs counter to prevailing public opinion, which favors life in a smaller-scale environment.[1] Correlaries of this trend toward bigness are the reduction of citizen participation in major events of importance, as personal time and energy resources restrict knowledge of the environment to the immediate and nearby; the maintenance of an illusion of participatory democracy, with elections intervening between crises; the failure of experts to comprehend the total urban environment, leading to fragmented and ineffectual decision making; and a dissolution of functioning urban government as public faith and management ability collapse in unison.

The appearance of professional urban management does not guarantee success in designing an urban environment with a desirable quality of life, as the SRI team observes:

> . . . the mere fact that a society can generate urban environments, and for a time adapt to them, does not necessarily ensure that it is a desirable thing to do. We can make errors and inadvertently accept living environments which may prove lethal to both our socio-psychological existence as beings seeking to unfold our potentials, and to our physical existence. . . .

> We are slow to search out or even recognize the existence of what may be . . . significant social crises . . . qualitative crises such as alienation, aesthetic deprivation, continuing mismatches between the character of the living environment and the altered need levels of people, overextension of social elasticity, etc. Unless a problem has a substantial physical manifestation which immediately threatens us, it is unlikely to get attention in an era plagued with crises. Yet the more subtle, qualitative problems, though they may manifest themselves only little by little in the short run, are of profound importance to our society.[2]

In his ground-breaking work, *The Image of the City*, Kevin Lynch notes that "a good environmental image gives its possessor an important sense of emotional security . . . the obverse of the fear that comes with disorientation."[3] Clifford Moller[4] and Felizitas Lenz-Romeiss[5] have expanded upon Lynch's theories, the former describing the important psychological effects of architectural space and the latter exploring affective patterns of spacial form and meaning.

But the occasions on which citizens are able to comprehend their physical environment, let alone the more complex and subtle sociopsychological environment of "community," are growing more infrequent. As the SRI study concluded, complexity and increasing scale of settlement are confounding citizens and experts alike. Now that the city's affairs are often also the affairs of the state and the federal governments, the boundaries of the city are growing more diffuse and unclear, and the relations within those boundaries are likewise becoming more difficult to clearly discern and understand.

The plight of the citizen is especially acute. In designing a citizen information system for Puerto Rico, John Little discovered that despite the professed need for channels of communication between citizens and officials, no organized, coordinated effort had been made in any city to provide these channels.[6] While citizen participation has been increasingly emphasized as a necessary component of sound government, citizens, in most cases, have been frustrated in their attempts at penetrating multilayered urban bureaucracy. Little credits this failure to the electronic media, which do a decent job of informing the citizen but which are incapable of aiding the citizen in informing the urban manager.

If an urban environment suitable for "unfolding our potentials" and enjoying life is to be realized, communication processes within the city must receive the attention of citizens and managers alike.

Until the role of urban communication is understood, there can be no scientific alteration of the process to produce a humane and socially progressive urban environment.

COMMUNICATION AND THE CITY

The human being is the center of the city, the primary client and sponsor of the city's activities. The informed collective decisions of human beings can lead to more effective management of the city; thus, it is important to understand how information reaches individuals and collective groups—how the city discourages, tolerates, or encourages rational, cooperative decision making among its inhabitants.

Richard Meier proposed the first formal theory of communication in the city: he hypothesized that cities are "agglomerations" of people seeking utility in transactions and that civic cohesion is a function of transactions in the public sector (as opposed to family, clan, or friendship group).[7] He further hypothesized that there is a need within citizens for a sense of personal control and social connection that must be satisfied for the city to thrive.

Meier drew upon information theory to analyze the "information flows" in the city, concluding that "communication stress" could result from information overloads affecting citizens in positions of responsibility and decision making. Meier's solution would be a greater societal reliance on automata for the transmission and presentation of data.[8]

Meier's pessimism regarding the inability of human beings to absorb information about their environment has been translated into a quasi-fascism in his more recent work. He suggests that

> as much as a quarter of the population may reject life styles based on an ever-accelerating volume of communications. These people, who strongly prefer a slower pace, will be willing to accept a much lower real income. They will be concentrated in such activities as the fine arts, personal service, growing perishable foods, part-time construction, and salvage. The "burned-out" cases will also have this supportive milieu to drop into, should they be forced to retire from most of their ties to the system.[9]

Informational misfits will be relegated to the "aging" inner city and "dying" resorts, while those able to accept responsibility will occupy the "living" suburbs. Finally, means will be established to protect those in "positions of responsibility" from "undue contact" with spurious and irrelevant information.

But this and similar structural remedies to presumed urban ills will always reflect the oppressive nature of the prevailing economic structure, unless basic tenets of the prevailing system are challenged and debunked.

For example, the Committee on Telecommunications of the National Academy of Engineering, which proclaimed that "the ultimate concern of any level of government . . . is the improvement of the quality of life for its people," added, "We advance the idea that telecommunications is a key resource, offering significant benefits in relation to its costs."[10] The Committee proposed an urban future in which the addition of telecommunications substantially upgraded the quality of many services currently supplied by cities, including public education and transportation (with cable channels substituting for physical rights-of-way).

However, the report of the Committee, like many of the other contemporary popular accounts of the "cable revolution," circumvented the difficult question of who or what would implement the cable systems envisioned. Perhaps, too much trust was placed in private entrepreneurs unable to live up to their own promises to serve the public interest. Harold Sackman has suggested that this trust was indeed misplaced:

> In a democracy, the public is the ultimate source of social power, and information power, accordingly, is ultimately a public trust. A public philosophy of information power needs to account for new democratic forms and procedures bearing on the organization and equitable distribution of social information.[11]

Seymour Mandlebaum specified a "philosophy of information power" to guide regulation of the cable medium, placing on the public agenda evaluation of publicly owned and operated channels of communication:

> . . . we should enter communications expenditures, with all the required alliances, in the lists of claimants together with school buildings, health clinics, roads, squad cars, and community centers. Instead of presuming that television policy deals with a home-studio system, dependent on private initiative and distinct from other information modes, we should be prepared to discuss publicly designed systems, linking diverse channels and functions in richly imagined, complex networks.[12]

Mandlebaum's connection of community with communication processes became the byword of a new school of communication policy makers, who adopted social, as well as economic, perspectives in their analyses of potential cable developments. Mandlebaum's faith in liberal elements of the society—educators, doctors, city managers, and others—as appropriate advocates for public ownership was obviously naive, but his general redirection of policy from vague presumptions of public gain to more concrete discussions of how development might be controlled was certainly commendable.

One student of Mandlebaum, who has become a major voice in communication planning, is Kas Kalba, whose strengthening of Mandlebaum's tentative theory began to provide municipal officials with new ways of evaluating communication resources: ". . . information can be seen as a settlement's underlying currency. What a city or region does with information—the circulation patterns it engenders, its information-processing and storage capabilities—is largely what the city or region is."[13] Yet, even Kalba's more sophisticated economic modeling is derived from a functionalist concern for "optimization" of telecommunications as an enterprise; his treatment of the social function of communications is as facile as is that of Mandlebaum. When Kalba speaks of "telecommunications innovations," he means those aspects of urban communication technologies that "will impact the conventional concerns of planners, from industrial location to improved delivery of social services."[14] But the crisis of the cities cannot be solved by merely making more efficient those processes that already exist within the city—since the crisis is a human one.

The widespread infusion of broadband communication systems—creating many more possibilities for the carriage to citizens of information—will have profound social effects. Will it help solve, or will it exacerbate, the urban crisis? Television may already have had dramatic psychological effects upon urban inhabitants' perception of their surrounding environment. Hidetoshi Kato believes that through the television medium, symbolic space supplements physical space for most city dwellers—thus making the cities psychologically habitable even though their physical density may exceed a "safe" level: the television world becomes a symbolic sanctuary.[15]

If Kato is right, and such unprecedented psychological effects are presently occurring, can there be massive social ramifications as a result of the compounding of individual experiential modification? If so, it would seem appropriate to return some determination of quantity and quality of media exposure (and information exposure in general) to the urban citizenry at the receiving end.

It is not inconceivable that with the widespread availability of cable communications, urban dwellers could program their own "right" mix of information diversity and quantity from neighborhood to neighborhood. The social psychological environment, composed of symbolic exchanges, could be altered by changing the patterns and content of those exchanges.

Even more fundamentally, urban telecommunications could significantly remake the socioeconomic environment of the cities. Melvin Webber has discussed possibilities for the diffusion of urban populations into more pleasant rural regions.[16] But he is quick to point out that such diffusion could result in the further segregation of oppressed economic classes; unable to afford access to expensive technologies, they could not leave the physical squalor and information-impoverished urban cores.

If urban telecommunications are to be used wisely, they most follow—or at least parallel—the development of social organizations able to provide conscious guidance for their implementation. A paradox inherent in the use of the new urban telecommunications technologies is illuminated by William Melody:

> In technologically advanced economies, timely information and an effective communications network can be assets of far greater value than those recorded on balance sheets.

Yet at the same time,

> new technologies can provide improved circumstances for the development of communication systems as tools of overt manipulation of behavior.[17]

If the technological capacity for improving urban telecommunications, and urban communication in general, is imminent, then the urban public must accept responsibility for its development and implementation. If the public defaults on this responsibility, then it will inherit a communication system with all the deficiencies of today's popular media, designed by self-interested corporate institutions.

Sackman calls for "deliberate, institutionalized, continuing public experimentation for public affairs," with every citizen having "to learn to think for himself as both subject and experimenter, with lifelong responsibility for social planning, selecting and implementing new experiments, evaluating social effectiveness, and applying the results."[18] He predicts that

> mass information utilities, linked to the real-time information bases in the public domain, could conceivably provide the leading instrumentality for the public to scan the social scene, identify problems, contribute to social control, and provide continuing feedback on the interplay of pluralistic social experimentation.[19]

Pluralism, of course, must be tempered by a concern for the ever-present danger of hegemonic dominance by the oppressive forces of the present society. But a synthesis of social change and technological innovation in the city is no doubt possible and desirable.

As one example of a necessary priority mandated by an aware public, it may be required that any telecommunication facility planned for the city must reach into, and serve the needs of, the inner city, where demands for services are quite severe; traditionally, suburban districts receive first consideration from profit-oriented vendors of all services.[20]

Little has proposed six criteria that should be applied to any citizen information system:

1. *Strong* enough to curb buck passing, cut unnecessary red tape, and redress legitimate grievances.
2. *Independent* enough to give the citizen and elected officials a real check on the bureaucracy.
3. *Informative for the citizen* as to what programs are available to serve him or to be influenced by him.
4. *Accessible* enough to give all citizens an opportunity to communicate with their government.

5. *Informative for government* as to what citizens need, what they are getting, and how they feel about various programs.
6. *Comprehensive* enough to improve procedures, services and programs throughout government.[21]

A more thorough plan for citizen information systems would have to include two further criteria: *open to constant criticism* (to promote ongoing improvement of the system) and *allowing for collective discussion and action* (to maintain wide distribution of awareness and controlling skills throughout the city).

Public ownership of urban telecommunication systems–technological infrastructures for the functioning citizen information systems–is essential to the realization of information processes in the city that would fulfill these criteria. Public ownership would make possible citizen management of urban telecommunication systems; given the establishment of appropriate constraints on the growth of public bureaucracies, democracy would become an active ingredient in the specification, implementation, and management of the city's communication resource. Ideally, an urban telecommunication system so operated would further strengthen democracy in the city by making possible universal access to information and communication facilities, thus reducing the need for centralized, hierarchical management of the city.* In effect, a publicly owned and operated telecommunication system–with today's technology, a cable communication system–would make possible electronic forums throughout the city and the restoration of the potency of the citizen.

ACCESS VERSUS OWNERSHIP: THE DANGEROUS DIVERSION

Many communication theoreticians, either from frustration or impatience, have abandoned positions they once held favoring public ownership of the communications media in the city–despite the historic failure of privately owned media to program in the public interest. They have, instead, adopted a line of

---

*A demonstration of the power of electronic media for stimulating collective citizen action occurred as part of a recent planning experiment in Berkeley, California. Using a remotely guided television camera, planners simulated tours through proposed future developments for viewing by citizens gathered in local auditoriums. "Eventually," reported one planner, "participants could take off on such miniature excursions without ever leaving their own homes since the pictures could be transmitted to television sets hooked into a community cable TV system" (*Human Behavior*, "The Ant's-Eye View of Architecture," *Human Behavior* 3, no. 8 [August 1974]: 48). The planners predicted that frequent public exposure would lead to better-informed decision making by citizens voting on proposed environmental alterations.

reasoning that idealistically separates ownership from access.

Edwin Parker, for example, has recently taken the position that making cable a "common carrier"–forcing the owner of the medium to relinquish some control over channel programming–is sufficient to ensure the public's interest in the system's operation:

> If control is vested in the hands of commercial operators, then the results could be as bad or worse than the society depicted by Orwell. The issue is *not* who owns or operates the hardware facilities. The issue is whether the facilities, however owned, are accessible for use by all members of the society.[22]

Clearly, access to the medium is indispensable to the preservation of the electronic commonweal. But access without the means of determining the surrounding context for programming, and without the means for promotion and production of programming and information services, is certainly not the ultimate answer to the public's need for new modes of communication.

Parker's confusion is evident: he only vaguely appreciates the essential impotence of the public when it relinquishes ownership–the source of absolute control–of its means of communication:

> It makes a great difference to . . . *economic development* . . . whether a communication network (or new modes on the network) are treated as essential infrastructure needed for development or whether it is treated as a consumption or a luxury item to be provided only after demand is great. In the latter case, development will be slow, uneven, and probably inequitably distributed (emphasis added).[23]

On the contrary, forms of economic development have consistently preceded and determined the growth of systems of communication. Fabian socialism in Britain encouraged the growth of a nonprofit broadcasting organization, with tinges of class elitism; fascist and communist dictatorships have produced communication systems that are strictly one-way and oppressive; and the capitalistic liberalism of America has spawned a communication industry, also one-way, which generates a welter of trivial "products" for consumption. It may be true that a communication "network" sustains the prevailing social relationships, but it is fallacious to suggest that its conception determines those relationships. The social and economic relationship inevitably shapes communication systems.

Communication attorney Anne Branscomb, proposing to open the cable industry to investment from telephone and broadcasting interests, has demonstrated how the idealism of the "access" argument can be applied to the development of cable:

> Ownership of the hardware may very well be more economically and technically feasible in a monopolistic entity—whether privately or publicly owned. Regulation of ownership and control of the hardware is not the most important priority. By far the most important priority is that all potential users have equitable access to the transmission capability. . . . more human communication needs will only be met through use of the available technology if more humans participate in the planning and funding of telecommunications systems.[24]

As Robert Jacobson subsequently observed, Branscomb, having "deprioritized" regulation of ownership and control of the "hardware," is at a loss to describe how "more humans" can participate in the planning and funding of telecommunications systems.[25]

Advocacy of historically unsupported and utopian access rights is not restricted to liberal theoreticians; it has found new adherents among federal politicians in positions of considerable power. The President's Cabinet Committee on Cable Communications, counseled by Clay T. Whitehead, former director of the Office of Telecommunications Policy, attacked the "overregulation" of cable, implying that a lessened public role in directing the development of telecommunication systems would spur the growth of the systems for the benefit of all.[26] While supporting the views of Branscomb and other liberal theoreticians, these political figures do not share the apologists' disregard for ownership of the hardware that makes more effective human communications a possibility; they are working quite hard at preserving the sanctity of ownership of the hardware.

Today's private cable operators, in their public comments, evince a strange combination of idealism and pragmatism. Their statements vacillate between righteous moral indignation at the thought of public ownership and contempt for the alleged inability of the cities to manage cable communications.

Sometimes, the message is strident. Spectrum Communications, Inc., addressed this "public service" message to the citizens of Chicopee, Massachusetts, concerning a referendum on public ownership under consideration in 1972:

> A special message to voters of Wards 7, 8, and 9 in Chicopee. The question of city-owned and operated cable TV appears on the November 7 ballot. We think city money can better be spent on the real necessities of government. Cable TV is a retail business and belongs in the hands of private enterprise. A "No" vote on the cable TV question in Chicopee on November 7 means a vote for communications freedom!
>
> Vote your choice, but vote . . .
>
> [LIVE TAG] A public service message, paid for by Spectrum Communications, Inc.[27]

A similar campaign of public service messages was waged by the National Cable Television Association in Columbia, Missouri; participating in the effort was a mysterious "fire drill" team dispatched to the beleaguered forces of private enterprise in that Midwestern city.* Aided by the major local newspaper and "freedom of the press" government-media separationists at the University of Missouri, the NCTA succeeded in reversing a popular mandate for public ownership during a referendum in February 1973.

Alfred Stern, president of Warner Cable, and a self-proclaimed leader in the fight for "cable reality," has made some particularly scathing comments about the ability of urban managers to run a cable system. In Stern's assessment,

> City officials, it seems to me are already so accustomed to hearing their own departmental leaders explain away the reasons for failure that a similar series of excuses concerning the breakdown of cable service or the low level of cable quality would not, I believe, cause much of a stir. . . . This hue and cry in a municipally owned system would most probably, I'm afraid, fall on the deaf ears of hardened public officials, officials for whom an unresponsive stance is practically a way of life—except in an election year.[28]

If, however, a *private* operator secures a franchise in the face of such venal opposition, the public interest is secured, since

> the very knowledge of having to answer to these local governing bodies—these issuers of franchises, these renewers of licenses, *these custodians of the public trust*—surely serves to guarantee that we will exercise our own internal controls in an attempt to avoid such confrontations (emphasis added).[29]

One part of the industry's strategy has been to denigrate the major metropolitan areas as potential markets for cable exploitation,[30] and to suggest that two-way and informational nonbroadcast services are simply not feasible in the current urban environment.[31] According to Paul Kagan, publisher of *Cablecast*, "The real purpose of providing cable service, in the larger metropolitan areas, from an entrepreneurial point of view, is to increase the channel choice and other new entertainment options."[32]

---

*David Foster, then president of the NCTA, disclosed the existence of this fire drill squad in an address before the Federal Communications Bar Association on March 15, 1973; yet, a subsequent request for further information directed to Brian Owens, Office of Government Relations, on March 23, 1974, netted only official ignorance. On April 25, 1974, Owens disclosed that "the person who worked on that is no longer with NCTA." The name of the "person" could not be found.

In contrast, public advocate Don LeDuc predicts that as the economic recession abates, the industry will embark on a "whirlwind" of franchise acquisitions and favorable renewals.[33] Dennis McAlpine, a Wall Street analyst of some eminence in the field of innovative technologies, had this to say in response to an inquiry concerning the health of the industry: "After reading your letter, I feel somewhat like the last of the Mohicans. Contrary to popular opinion, much of the CATV industry is alive and well. In fact, the CATV stocks have done admirably this year (1975), being up about 180% from their year end close."[34]

If the cities do seek communication channels unencumbered with obligations to external economic interests, they will not find a solution in reformist legislation protecting a theoretical "access"; the history of such legislation is replete with instances of inadequate enforcement and distorted application. Ownership of an urban telecommunications system by the public, guaranteeing operations of the system in the interests of that public, would be a more certain answer to the cities' needs.

THE ENTREPRENEURIAL IDEALISM OF COMMUNITY OWNERSHIP

Proponents of "community ownership" for cable communications occupy a strange middle ground between the forces of private enterprise and the advocates of public ownership for urban media. On the one hand, they wish to keep economic control of cable in the community it serves; on the other, they fear the concept of municipal ownership as an attack by city hall on local autonomy. Since a great number of those supporting community ownership—direct investment of capital in locally controlled enterprise—come from minority communities, which are often the victims of municipal abuse or neglect, this lack of faith in municipal ownership is understandable.

Too often, however, this antagonism to public ownership accompanies a cynical inability to conceive of radical alterations in the structure and operations of municipal government—even though it appears that the crisis of the cities has softened resistance to such alterations in many cities. Further, the advocates of community ownership tend to uphold the prevailing capitalist ethic, operating as traditional entrepreneurs attempting to exploit a new market and distancing themselves from the less affluent of their communities.

Charles Tates is perhaps the most forceful advocate of community ownership; formerly with the Urban Institute, he authored a monograph suggesting that direct ownership of communications media, at least in minority communities, should remain in the hands of local residents. Tate feared that while multiple-system operators (MSOs) and other white-dominated owners would be totally unresponsive to the minority community needs, municipal managers could exercise a more insidious oppression:

> Local governments, power blocs and the white electorate will probably resist minority ownership and control of CATV systems under any circumstances. Municipal ownership could be a convenient mechanism for maintaining white control. Blacks are woefully under-represented in the policy-making structure of existing municipally-controlled agencies and corporations such as public housing, port, and airport authorities. Unless there are unusual legal proscriptions that favor and protect minority control of community systems, similar conditions of exclusion and under-representation are likely to emerge in the policy-making bodies of municipally-owned systems.[35]

Tate acknowledges one major drawback to his concept of community ownership: venders probably will not want to underwrite an expensive project serving the poorest elements of the urban society. In some desperation, Tate must rely upon the very city government he deprecates as the white "power bloc":

> Promotion costs may be higher in the first three years in the inner city [as well as interest costs on low-equity loans]. Some of these costs could be absorbed by the city government, community development corporations, model cities programs, civic and social organizations. For example, the city could allocate funds for a promotion campaign against future receipts from franchise fees.[36]

As Tate chooses to work in the prevailing system, he is doomed to suffer its inequities. Few minority entrepreneurs have secured funding for their projects; after five years of struggle the Mafundi Institute's cable project in the Watts community of Los Angeles is extinct due to lack of funds, as is the Venice Community Cable Corporation in that developer-threatened part of Los Angeles.

One solution to this dilemma, recommended by Lois Brown, cofounder of the Philadelphia Cable Coalition, is to invite multiple-system operators to finance the development of the large-market system, on condition that community participation in the venture be formally stipulated in any contract between the MSO and the city.[37] "Of course," admits Brown, "many problems exist with partnerships, including the fact that I've come to the conclusion that most of the MSO's are less than admirable."[38]

Neither Tate nor Brown chooses to confront the other obvious defects in community ownership: the possibility that community representatives will be coopted, through financial enticements, to cooperate with MSO management or other large investors; the likelihood that ownership of the system will, if the system is successful, tend to leave the local community through sale to outside buyers; the difficulty of eventually linking systems that are dissimilar in technology from community to community; and perhaps most significantly, the virtual certainty that even a locally owned system will be dominated by oppressive interests within the local community—doctors, lawyers, church leaders, and other affluent elements of the community.

Community ownership may be a cure for racism, but it only sustains a local social relationship that mirrors the external society: the dominant classes, controlling the system, vitiate any potential the technology might possess for hastening a social transformation leading to an improvement of the quality of life for all inhabitants of the community.

A better alternative to local entrepreneurial enterprises might be to have the municipal government fund and supervise development of a comprehensive, citywide cable system, and then to have the municipal entity either segment the system into community-operated units or sell the parts of the system to local community trusts established with public funds. No minority entrepreneurs will make windfall profits under this arrangement, but it has greater potential for community control than reliance on traditional market tactics of the exploitive private sector.

OBSERVING THE STATUS QUO

It is necessary to confront the staunchest arguments of the private enterprise opposition and, in the tradition of critical thinking, to allow the contradictions in the logic of the "rugged individualists"—actually, corporate managers—to become self-evident.

There *are* contradictions. They are manifested as private enterprise appeals for deregulations, on the one hand, and, on the other, for legislation prohibiting municipal ownership of the media.* Most decisively, these contradictions exist in the historical process of cable's development in this country, a process that has united the supposed protectors of the public trust, the Federal Communications Commission, with the cable industry in a conscious effort to transform cable from a beneficial technology for all city dwellers to a luxury, superficial "box office" for an affluent few.

---

*Senate Joint Resolution 125, introduced in the Senate as a joint resolution with the House. The bill died with the Ninety-Third Congress in 1973, but it has been alluded to in informal conversations between the author and members of the cable industry as a veiled threat.

Anaheim's Mayor Jack C. Dutton has pointed out the inconsistencies in the cable industry's position, adding that "municipally-owned systems have historically been provided only when independent entrepreneurs have refused or been unable to provide such service" (U.S. Conference of Mayors, *Mayors' Action Report on Cable Television* [Washington, D.C.: U.S. Conference of Mayors, 1974], p. 4).

This process, the degradation of cable service over the last five years, might be the best argument of all for the public's assumption of ownership and control of urban telecommunications media.

NOTES

1. Duane Elgin et al., *City Size and the Quality of Life* (Washington, D.C.: National Science Foundation, 1974), p. 2.

2. Ibid., pp. 60–62.

3. Kevin Lynch, *The Image of the City* (Cambridge, Mass.: MIT Press, 1960), pp. 4–5.

4. Clifford B. Moller, *Architectural Environment and Our Mental Health* (New York: Horizon Press, 1968).

5. Felizitas Lenz-Romeiss, *The City–New Town or Home Town?* (New York: Praeger, 1973).

6. John C. Little, Chandler A. Stevens, and Peter Tropp, "Puerto Rico's Citizen Feedback System," in *Analysis of Public Systems*, ed. Alvin W. Drake, Ralph L. Kinney, and Philip M. Morse (Cambridge, Mass.: MIT Press, 1972), pp. 432–35.

7. Richard L. Meier, *A Communications Theory of Urban Change* (Cambridge, Mass.: MIT Press, 1962), pp. 44–45.

8. Ibid., pp. 18–19.

9. Richard L. Meier, "Urban Ecostructures in a Cybernetic Age: Responses to Communications Stress," in *Communications Technology and Social Policy*, ed. George Gerbner, Larry P. Gross, and William H. Melody (New York: John Wiley & Sons, 1973), p. 361.

10. National Academy of Engineering, Committee on Telecommunications, *Telecommunications for Enhanced Metropolitan Function and Form* (Washington, D.C.: National Academy of Engineering, 1969), p. 7.

11. Harold Sackman, *Mass Information Utilities and Social Excellence* (Princeton, N.J.: Auerbach, 1971), p. 248.

12. Seymour J. Mandlebaum, *Community and Communications* (New York: W. W. Norton, 1972), p. 120.

13. Kas Kalba, "Telecommunications for Future Human Settlements: A Planning Framework for Minnesota Experimental City," *Ekistics* 35, no. 211 (June 1973): 329.

14. Kas Kalba, "Urban Telecommunications: A New Planning Context," *Socio-Economic Planning Science* 8 (February 1974): 40.

15. Hidetoshi Kato, "From Pantheon to Presley: Changes in Urban Symbolism," in *Communication and the City: The Changing Environment*, ed. East-West Communication Institute (Honolulu: East-West Communication Institute, 1973), paper no. 7, p. 15.

16. Melvin M. Webber, "Urbanization and Communications," in *Communications Technology*, ed. Gerbner, Gross, and Melody, pp. 297–303.

17. William H. Melody, "The Role of Advocacy in Public Policy Planning," in *Communications Technology*, ed. Gerbner, Gross, and Melody, p. 166.

18. Sackman, *Mass Information Utilities*, p. 256.

19. Ibid.

20. Alan K. Campbell and Seymour Sacks, *Metropolitan America* (New York: The Free Press, 1967).

21. Little et al., "Puerto Rico's Citizen," pp. 118–19.

22. Edwin B. Parker, "Networks for an Information Society," *Bulletin of the American Society for Information Science* 2, no. 1 (June-July 1975): 13.

23. Ibid., p. 13.

24. Anne Branscomb, "Killing the Goose that Lays the Golden Egg," *Journal of Communication* 25, no. 3 (summer 1975): 221-22.

25. Robert E. Jacobson, "'The Cable Fable': Lost in Valhalla," *Journal of Communication* 25, no. 3 (summer 1975): 219-21.

26. President's Cabinet Committee on Cable Communications, *Cable Report* to the President, January 14, 1974, reproduced in *Television Digest* (January 21, 1974), special insert.

27. Spectrum Communications, Inc., Radio commercial preceding November 7, 1972, referendum; Chicopee, Mass.; obtained from National Cable Television Association, Washington, D.C.

28. Alfred R. Stern, Speech before the New England Cable Television Association (Hyannis, Mass., April 18, 1973), p. 3.

29. Ibid.

30. "In Cable Franchising, the Cards Are Being Played Closer to the Vest," *Broadcasting* (February 4, 1974): 48-49.

31. "Cable-TV Dangles New Lures," *Broadcasting* (December 1, 1973): 61-64.

32. MITRE Corporation, *Symposium on Urban Cable Television*, vol. 2 (McLean, Va.: MITRE Corporation, 1973), p. 7.

33. Don R. LeDuc, "Cable-TV Franchising and the Local Bar Association," *Wisconsin Bar Bulletin* 47, no. 6 (December 1974): 41.

34. Dennis McAlpine, of Anthony Tucker and R. L. Day, New York, private communication, May 14, 1975.

35. Charles Tate, *Cable Television in the Cities* (Washington, D.C.: The Urban Institute, 1971), p. 36.

36. Ibid., p. 27.

37. Lois Brown, of Metrotel Communications, Philadelphia, private communication, February 17, 1975.

38. Ibid.

CHAPTER

4

TELECOMMUNICATIONS IN THE URBAN CONTEXT

THE PUBLIC INFORMATION VACUUM

Ideological opposition to government involvement with communications media remains a powerful obstacle to public support of comprehensive urban telecommunications systems. Undoubtedly, responsibility for the persistance of this ideology among many Americans properly rests with the corporate owners of most of today's media.

There is no "free press" in contemporary America. Citizens may secure access to the broadcasting media only at the behest of the networks and independent owners, who hold their licenses in virtual perpetuity. Access to the telephone system is "free," in the sense of its being available for home use without direct external control of that availability, but this freedom is purchased at the price of tremendous profits and political influence for the American Telephone and Telegraph Company (AT&T), General Telephone and Electric Company (GTE), and a handful of smaller firms. For all but the smallest handful of private citizens and public institutions, satellite-aided communication—free of intermediary commercial "brokers"—remains a distant dream.

An unqualified defense of the "free press" under these conditions is protective of corporate interests, which misuse the public trust granted to them; it is a dangerous position, willfully ignorant of the social dysfunctions (for example, overconsumption, public apathy, vigilantism, and so forth) engendered and sustained by privately controlled communication media—individualistic, alienating, and confounding—supportive of the prevailing hegemony.[1]

Yet, the rather remarkable perversion of the First Amendment necessary to bring about this state of affairs continues to be seen as "traditional" and "right," even by experienced communication analysts and commentators:

> As commercial cable development of two-way service has faltered, a few observers have called for government construction and ownership of cable television system altogether. They contend that communications should be considered as important for tax support as roads, sewers, and water lines. Although government-owned communications systems are the rule in other parts of the world, the U.S. Constitutional guarantee against government control of the media makes this proposal untenable to most Americans.[2]

Encouraging communication planners to adopt an "advocacy" role on behalf of the unorganized public's interest, William Melody correctly identifies the principal activity of the corporate communication interests: "The full economic and political power of the communications industries is clearly used to insulate, preserve, and extend their markets, as well as to control the development and application of new technologies. . . ."[3]

In working to extend their markets, the industries foster the notion that they are fulfilling their obligation to serve the public. For example, broadcasters often claim, according to Richard Stavins, that

> TV is democratic and has discharged its fiduciary function because it gives the public what the public wants. This claim, though seemingly forthright, is in fact the most insidious piece of indoctrination yet invented by television: a verbal trump card placed in the hand of every licensee. . . . Were we to inquire of the industry itself how the public interest could possibly be served through a steady diet of moribund, petrified, canned re-runs, with only an occasional helping of mental nourishment, we are immediately presented with the trump card.[4]

According to a survey conducted by Stavins, broadcasters serve the public interest by contributing only 2.5 percent of all programming time to public affairs programming.

The public, consuming the trivial and redundant information supplied by broadcasters in the form of situation comedies, sports coverage, and violent police shows, is not, as these industries claim, free to pick and choose. Alexander Tzonis has synthesized a rationale for the consuming public's willingness to settle for the mediocre, and even the destructive, in the marketplace:

> The consumer needs . . . products not because this is his nature but because he is disciplined to do so. . . . The need of the consumer has been determined in such a manner as to be coordinated with the potential of the supplier of the commodity. . . .
>
> But the oppression which man suffers by this transaction is certainly greater than his gratification from the fetish. Not only is his freedom

> curtailed by the dependence on the supplier, but also the possibilities of overcoming oppression in his association with other human beings . . . are limited, since by every transaction he is strengthening the status quo as a whole.[5]

Tzonis's analysis is applicable to the ephemeral but socially damaging product of the conventional communication media, structured to achieve public apathy and private exploitation.

Jerrold Oppenheim, reviewing the development of cable in the United States, has lamented the abandonment of the medium to the marketplace; it has given the industry license to determine an appropriate level of service to be provided the public–

> and what has it come up with? What has it, in its profit-seeking wisdom, decided the most important first function should be of the most important technological advance in communications since the telephone?
>
> Movies. Not-quite-first-run feature films on a premium payment basis. The industry is virtually united: the only way cable will sell in big cities is if it offers a new service. . . . Sell us your pictures, moviemen, and we will profit by hitching pay-TV boxes to home television sets.[6]

Melody has speculated that the cable industry will develop along the lines of the oligopolistic telephone industry in order to gain greater financial leverage and to conceal internal decision making behind corporate anonymity; if no one intervenes, Melody foresees the largest MSO, TelePrompter Corporation, becoming "a resident East India Company."[7] As large corporations beholden to a few nonresident investors, cable operators will necessarily devote fewer resources to local needs, unless there are significant and immediate profits involved.[8]

Cable is not likely soon to adequately service most of America's large cities.[9] Cable operators are beginning to recover acceptable rates of return in those areas already cabled, the most affluent and approachable for the marketing of premium programming. There is no incentive for the operator to risk developing the less affluent sections of the city, to build inward from the suburbs; such a move might endanger the profits now finally being realized with the reversal of cash flow from "out" (construction, staffing-up, and so forth) to "in" (monthly fees, rental of selector boxes, subscriptions to premium services, and so forth).

Keeping communication in the marketplace will certainly enforce the prevailing status quo separating privileged from underprivileged; keeping urban telecommunications out of the public sector will accelerate class divisions and urban disunity, which are at the roots of today's crisis of the city.

(Citing "the overwhelming weight of evidence for the public interest in a free communications system," the American Civil Liberties Union has called for "the adoption of an open, two-way "telephone" [-like cable] system, established, regulated, and, if necessary, financed by the government."[10])

In the absence of a national ethic encouraging the growth of public communications media, and, especially of *publicly accessible* media—and given the apparent desire of cable operators to simply emulate the programming content and policies of conventional television—there is little reason to cheer the end of information scarcity, as more idealistic observers have done.[11]

FILLING THE VOID

The process of communication itself is intangible: it is impossible to identify the moment in space and time at which one human being's imagery is replicated in the mind of one or more others. But the effects of that process, and the long-term implications for the society in which it occurs, can be profound indeed. Natan Katzman has proposed a number of "theoretical premises" related to the introduction of communication technologies in complex social environments.[12] Conceding that the new technologies increase the amount of information reaching each individual, Katzman is quick to qualify this concession: gaps that exist in information usage, between the information-rich and the information-poor, are enlarged by the new technology. The information-rich—usually the affluent and socially sophisticated—gain in information utility at a much greater rate than the information-poor, who are usually the economically powerless.

Katzman agrees with Meier's prediction that as the quantity of information reaching individuals exceeds their ability to comprehend and utilize it, machines will allow human beings to summarize and store information for optimal use. Unlike Meier, however, Katzman is aware of class-based imbalances in access to the machines (though he prefers to speak in terms of advantaged and disadvantaged "groups"). Rather than settling for Meier's resigned structuralist solution to technology-induced dysfunctions, Katzman recommends an end to free market determination of when and how technologies will arrive. The role of technologies in society is a decision that belongs in the public forum.

As crucial as a planned program for technological use is to the socially progressive, information-gap-narrowing effort, Katzman believes that positive action must also be taken to ensure that all social strata have motivation for the use of, as well as access to, the new technologies. What is needed is a means for locating and correcting those social conditions that prematurely destroy motivation; unfortunately, new technologies come along to widen existing information gaps or create new ones, at an increasingly rapid pace.

Only simultaneous programs of social reorganization and the implementation of progressively designed technologies—technologies designed to increase public access and stimulate collective activation—can expect success: effective organization must create a public willing to guide the development of the technology and take responsibility for it, while the technology must provide a focus for action and provide unfiltered channels of communication that are conducive to organization and action to satisfy common needs.

Such a strategy has been adopted by the Canadian National Film Board's "Challenge for Change" program.[13] Equipped first with film, and later with portable video equipment, teams of "facilitators" enter rural and urban communities and use these technologies to identify local problems and assist in building community coalitions to correct them. Never completely able to overcome powerful institutions resisting change—local administrators, absentee landlords, even the press—these facilitators have been further hampered by a lack of accessible media able to convey information beyond the immediate environment of a screening room or street corner show. When they have the cooperation of local cable operators, the facilitators enthuse about their success in letters and at informal gatherings of video enthusiasts; when that cooperation is not forthcoming, their efforts are often a source of great disappointment.

As a result, Canadian groups still working for social change, such as the Communication Research Action Project at Simon Fraser University, have chosen to pursue on behalf of their clients the dual goals of access to the media and local ownership.[14] For example, in providing radio links for coastal Indian tribes in British Columbia, the Simon Fraser activists worked to secure ownership of the system hardware for the Indians; this ensured continued access to the medium in accordance with local needs, and it demanded of the Indians the acquisition of skills for the maintenance and management of the sophisticated radio technology. This kind of learning will benefit the Indians in their dealings with other Canadians, while the radio links will serve, in a real sense, as technological support for collective organization. The possibility of rapid passage of information, combined with mass organization around collective goals, increases the possibility for eventual success of the reestablished coastal community; total involvement of the community guarantees that the technology will always be democratically used for collective purposes.

In the province of Saskatchewan, the provincial government has made plans to acquire all existing cable systems in the province and to extend cable service throughout the province.[15] The aim is not manifestly progressive in its present context: the provincial government is more intent on strengthening its power in remote northern regions than in transforming the social order to facilitate democratic decision making involving all citizens. Nonetheless, the provincial cable system will allow communities and individuals existing in virtual isolation to learn of the concerns of others throughout the province, possibly leading to collective action that would be self-sustaining and accelerative in nature.

Edwin Parker has suggested the use of cable-linked information systems for the organization of disadvantaged groups. There is a geometric increase in involvement made possible by the progressive use of new information technologies:

> One potential feature of the new information technology . . . will be the capacity to use the information system to locate other people with similar interests, so that it can be used as a tool to aid in mobilizing manpower, on short notice, if necessary, to counter attempts by existing institutions to use their technologies for their own benefit when that benefit is at the expense of the rest of the people in the society.[16]

Parker's reformist tendency diminishes the scope of his recommendation: his vision is one of groups organizing to participate in formal technology assessment—a limited action at best. Nevertheless, Parker has helped to define a useful role for urban telecommunications—bringing information resources to the working classes, the uninformed majority of citizens in urban society.

The assumption of this responsibility by the city cannot come too soon. Citizen movements are still resource starved, but given the potential for establishing more comprehensive information-receiving and information-disseminating communication links, these limited movements could prove to be prototypical of more widespread urban activism.[17] In an urban setting, with communication links in place—and with a general willingness among the population to reorder existing social relations—isolated citizen movements and organizing efforts could coalesce into a genuine popular movement in the city.

Responding to the social crisis of the cities—alienation among their inhabitants, disintegration of stabilizing neighborhoods, and the flight of the affluent to the suburbs—citizens could express their need for greater information resources by supporting the construction of a citywide, publicly owned, publicly accessible communication infrastructure.

CABLE COMMUNICATIONS AS PUBLIC ENTERPRISE

Private development of cable television has been less than energetic: not only has the spread of the actual "hardware" been curtailed to preserve waning capital resources but the services offered to an affluent few have been of the most trivial variety. So it should not be surprising to learn that cable technology has atrophied along with the grandiose visions of self-interested entrepreneurs, whose sense of public duty evaporated at the first hint of public resistance to their incursions.

In the absence of private initiatives leading to widespread local availability of cable, and with the abdication of the FCC from responsibility for the protection of the public interest, only the cities can secure the public's right–and need–for plentiful and unfiltered two-way channels of communication.

Public support of the media of communication violates no canon of American law or custom (as indicated by Congressional support of public broadcasting to the tune of $500 million in 1976). But governments at all levels have been discouraged from investing in the development of cable communications by those who would suffer most from such an undertaking–those who have local or national monopolies in existing media.

Political usurpation of the public's communications resources has often been held forth as a potential threat in order to deter any consideration of public ownership. However, this is hardly a compelling argument against politicization of corporate enterprises (public and private). As Peter Barnes and Derek Shearer have observed, in examining the worth of public enterprises, "The objection that public enterprise might become too political rings particularly hollow when one considers the many ways in which private corporations have used politics and politicians and have benefitted from public subsidies."[18] Edward Roth has also injected a note of sensibility in the sometimes shrill debate over "government control of communications," claiming that even within the present social structure, "it is unfair to assume that city officials are any less interested than others in protecting the Constitutional rights of citizens."[19]

Harold Sackman considers the question of control versus personal freedom–a freedom that is becoming quite circumscribed in the contemporary city–from another perspective. His concern is for the collective morale and the growth of community, rather than with the narrowing of personal experience and withdrawal. Without explicitly advocating public investment in communications media, it is clear that Sackman would support the development of a communications medium under public stewardship–a medium that could, finally, contribute to "social excellence":

> For many, the emotional appeal of privacy is linked to the preservation of individuality and personal freedom. The thinking goes something like this: the more others know about me, the more vulnerable will I become, and the more harm they can inflict on me. Anonymity is equated with freedom. But the argument cuts both ways. The more an individual knows, the more options he has, the richer his individuality, and the greater his chances for creative growth. Greater sharing of knowledge is here equated with rich social communion and greater creative freedom.[20]

Whether one adopts a legal or social stance in evaluating the benefits that could be provided for the public by its own cable communications system, it becomes very clear that arguments offered against public investment are ideologically transparent–and that the only substantive arguments against public cable communications are pragmatic, relying upon public apathy and conditioned resistance to public enterprise for their validity. Once that validity is contradicted by the successful reality of public cable—as it one day must be—Sackman's hoped-for renaissance of information-linked community will become possible in the city.

NOTES

1. Herbert I. Schiller, *The Mind Managers* (Boston: Beacon, 1973).

2. Walter S. Baer, *Cable Television in the United States—Revolution or Evolution?* (Santa Monica, Calif.: Rand Corporation, 1974), p. 12.

3. William H. Melody, "The Role of Advocacy in Public Policy Planning," in *Communications Technology and Social Policy*, ed. George Gerbner, Larry P. Gross, and William H. Melody (New York: John Wiley & Sons, 1973), p. 177.

4. Richard L. Stavins, ed., *Television Today* (Washington, D.C.: Communication Service Corp., 1969), p. 72.

5. Alexander Tzonis, *Towards a Non-Oppressive Environment* (Boston: i press, 1972), pp. 104–05.

6. Jerrold N. Oppenheim, "The Unfulfilled Promise of Cable TV," *The Progressive* 38, no. 2 (February 1974): 49–50.

7. Melody, "The Role of Advocacy," pp. 176–77.

8. Robin Marris, "Businesses, Economics, and Society," in *Social Innovation in the City*, ed. Richard S. Rosenbloom and Robin Marris (Cambridge, Mass.: Harvard University Press, 1969), p. 30.

9. Dennis McAlpine, of Anthony Tucker and R. L. Day, New York, private communication, May 14, 1975.

10. American Civil Liberties Union, *Policy #20, Cable Television* (New York: ACLU, July 22, 1968).

11. Ben Bagdikian, *The Information Machines* (New York: Harper & Row, 1971).

12. Natan Katzman, "The Impact of Communication Technology," *Ekistics* 225 (August 1974): 125–30.

13. Dorothy Todd Hénaut, "A Few Notes on Regional Projects," *Access, Challenge for Change/Société Nouvelle*, no. 10 (Autumn 1972): 11.

14. Gail Martin, of Communication Research Action Project, Simon Fraser University, Burnaby, British Columbia, Canada, private communication, August 20, 1975.

15. G. D. McCormick, of Saskatchewan Telecommunications, Regina, Saskatchewan, private communication, October 23, 1975.

16. Edwin B. Barker, "Technology Assessment or Institutional Change?" in *Communications Technology*, ed. Gerbner, Gross, and Melody, p. 544.

17. Hazel Henderson, "Information and the New Movements for Citizen Participation," *The Annals of the American Academy of Political and Social Science* 412 (March 1974): 34–43.

18. Peter Barnes and Derek Shearer, "Beyond Antitrust," *The New Republic* 172, nos. 1–2 (July 6 and 13, 1974): 18–19.

19. Edward J. Roth, "Should City Hall Own the Cable TV System?" *Nation's Cities* 11, no. 6 (June 1973): 42.

20. Harold Sackman, *Mass Information Utilities and Social Excellence* (Princeton, N.J.: Auerbach, 1971), p. 209.

CHAPTER

5

THE FAILURE OF "FREE ENTERPRISE"

WHO PROTECTS THE PUBLIC?

When the California Council on Intergovernmental Relations published a summary of local government planning activities in California in 1972, it exposed the parochial characteristics of most urban planning in city government.[1] Among the major concerns of the smaller cities, revenue management was preeminent; among the larger cities, environmental quality (especially with the threat of federal enforcement) received the most attention. In a list of urban concerns—including revenues, land use, environmental quality, housing, public facilities, transportation, and open space—urban communication did not even qualify as a topic for consideration.

California is the most heavily cabled state in the country, but municipal managers involved in urban planning have not awakened to the presence of this unknown quantity—cable television—within their own cities' boundaries.

California is not unique in its "cable illiteracy." The vast majority of states have yet to authorize any formal cable responsibility at either the state or the local level, resulting in a regulatory vacuum that leaves the public interest in cable communications undefined and unprotected. Richard Leone and Robert Powell, after conducting a survey of New Jersey municipal franchising activities, concluded:

> Local governing bodies move quickly on cable largely because they do not recognize the policy issues posed by this medium. As a former solicitor for a southern New Jersey town told us, "No one in this town had ever heard of cable television before the company came to us for a franchise. No one presented any argument against awarding the franchise and the company said that the town would

> get some money from the system once it was in operation. What else was there to consider?"[2]

Ralph Lee Smith has described a novel marketing technique employed by aggressive cable operators, "Rent a Civic Leader": stock options are traded to local influential persons in return for public support of the franchise.[3] In Philadelphia, a citizen group has been contesting the trading of franchises among large multiple-system operators.[4]

Apparently, municipal franchising authorities are in little better shape to deal with large cable operators than they were two and three years ago. (The city of Los Angeles, after more than a year of policy planning, accomplished little more than renaming the city's franchises with its largest operators.) Don LeDuc, advocating involvement of the local bar in franchising activities, has warned that "a city unprepared for a whirlwind promotional campaign by a skilled national organization may well find itself unable in the future to resist accepting a cable industry prepared franchise."[5]

LeDuc is equally critical of the role of the Federal Communications Commission; in a comprehensive study of the FCC's regulation of communication technologies, LeDuc found the FCC to be "simply a supervisor of industry activities . . . capable of innovation only to the extent that the interests of the industry and the public seem to coincide."[6] He observed correctly that where nonbroadcast programming has grown, it has grown under the stewardship of local agencies; according to LeDuc, the FCC is disinterested in any aspect of cable not impinging on the profitable operations of conventional broadcasting.

There is no historical reason to presume that the FCC is prepared to defend the public interest in the absence of strong local initiatives. The agency only recently took a neutral position in the matter of "two-tier" versus "three-tier" regulation of cable; previously, the FCC had advocated the surrender of either the local or the state authority as intermediary regulator of cable activities, thus reducing the obligations of cable operators from three to two agencies (a local agency and the FCC).[7] Had the states been forced to choose between securing their own powers and empowering the cities to regulate cable formally, all local prerogatives might have been removed to central (and industry-accessible) state capitals, hidden safely away from the citizens. Fortunately, neither the cities nor the states have capitulated to this drive toward centralization of communications management.

Other critics have lamented the inability of the FCC to even enforce its own written regulations. Jerrold Oppenheim has documented the frequent issuance by the FCC of waivers to operators unwilling to comply with the regulations for minimum service.[8] In one case, an operator serving five communities (totaling 2,000 inhabitants) was granted a waiver allowing the provision of one set of access channels–municipal, educational, and public–for all five systems. A larger company then requested a similar waiver for four of its systems, serving

12,000 inhabitants, to which the FCC agreed, citing the earlier precedent.

In the same expose, Oppenheim has described occasions when the FCC granted certificates-of-compliance to operators whose franchises had already been revoked by local authorities. (The author of this book, while investigating the records of Theta Cable of California, discovered letters from Theta's Washington attorney to the FCC's Cable Television Bureau, asking for rapid approval of the franchise for El Segundo, California, even though no provisions existed there for public access or other forms of local programming. The major cause for haste, according to Theta, was the need to experiment with two-way services in that city—services that are now defunct only three years later. The certificate-of-compliance was awarded in a record 30 days, apparently without public hearing.)

The FCC's abdication of regulatory responsibility has rapidly accelerated in 1976-77. In 1972, the agency had already exempted from compliance with its 1972 rules all cable systems in operation prior to 1972. These exempted systems comprise 98.5 percent of all operating cable systems. But still in force was the formal intention of the FCC to "review" those rules in 1977, requiring all systems to provide, by that time, two-way capability, access channels, and leased channels in all major markets. In response, the National Cable Television Association loudly proclaimed, in 1974, that origination efforts—programming provided by the local operator independent of television broadcasting—had jumped 14 percent, with the increased number of systems serving 24 percent more subscribers.[9]

Such local origination and access programming did not often meet local programming needs;[10] expectedly, most private money had been invested in operator-controlled programming, and, especially, pay-movies and pay-sports, which did not demand any involvement in local affairs.

Then, the whole question of local origination and public access programming became moot when the FCC issued its opinion on Docket no. 19988, FCC 74-1278, 49 FCC 2d 1078 (1974). With this ruling, the FCC ended all local origination requirements, apparently at the urging of the NCTA, which, only a few months before, had been boasting of a "14% jump" in local origination efforts.[11] At the same time, the FCC decided to formalize its requirements for public access—but only for systems with more than 3,500 subscribers (again, not the bulk of the systems). As a result, a large local system must now keep on hand equipment with "a cost less than $10,000, with equipment maintenance costs expected to be less than $1,000 per year,"—or about $2.75 per subscriber each year. It is difficult to foresee what programming of significance can be produced and promoted with these constraints in force.

Acting in an industry-protective role, the FCC warned "other authorities" to avoid imposing their own local origination requirements, supposedly because such efforts might demonstrate the "same infirmities we found in our attempt at mandating programming."[12]

It was only six months later that the FCC, possibly goaded by the Nixon White House Office of Telecommunication Policy's own efforts to have cable deregulated,[13] began the near total abandonment of its regulatory role promoting innovative uses of cable television. With an opinion on Docket no. 20363, FCC 75-821, 54 FCC 2d 207 (1975) the FCC released all exempted cable systems—those that had been "grandfathered" out of compliance with the 1972 rules—from the 1977 deadline for compliance. All of the access, leased-channel, and two-way promises of cable evaporated in a moment. In the face of extremely heavy lobbying by cable interests, and despite criticism leveled at industry claims by the Cable Television Information Center and others, the FCC had decided that the industry simply could not afford the costs of construction necessary to bring their systems into compliance.

> Joint comments filed on behalf of 22 cable operators contended that at present there has been no demonstration of a need for two-way capability of equivalent bandwidth for nonbroadcast purposes. Furthermore, they said, expanding channel capacity from 12 to 20 would result only in more unused channels, since "most" major market systems are now carrying all the channels permitted by the FCC rules.[14]

The FCC did continue to call for compliance on the part of newly franchised systems. But since these comprised only a small portion of the total industry, the FCC had, in effect, liberated the industry from any sort of obligation to experiment with local programming or two-way information services.*

The large corporations, while consolidating their control of cable communications through the development of satellite "networks" and in pay-movie syndication, can somehow no longer find the money for experimentation in the area of public services:

> What is taking place is a "return to reality." That was the expression used by Warner Cable Chairman Alfred Stern in a December speech before a convention of municipal officials, in which he pointed out the need for both the cities and the industry to discard their "blue sky" assessments of cable and concentrate on the medium's present capabilities.[15]

Consequently, MSOs have begun to exert pressure against municipalities and other clients that do not toe the industry line. In one town, TelePrompter

---

*The FCC moved even further toward total deregulation of cable by opening Docket no. 21002, FCC 76-1070 (released December 9, 1976) in response to petitions from cable industry associations requesting review and removal of many of the 1972 rules.

Corporation, largest cable operator, "cut back" service because the town granted a rate increase 25 cents per month less than surrounding communities on the same system.[16] In Oakland, Teleprompter's Focus Cable subsidiary has yet to go into full-fledged operation despite extravagant claims made on the part of the operator when the franchise was secured; plans for two-way and local origination facilities have been curtailed pending further developments.[17] And, in Southern California, a Times-Mirror Corporation cable system brought the San Clemente City Council to court for failure to approve a "required" rate increase; the corporation's case was dismissed in court, although an appeal was promised.

Finally, in Tallahassee, Florida, Clearview-Westinghouse Cable discontinued its public access efforts, leading local community groups to proclaim:

> In light of the Clearview-Westinghouse perception of its role in serving the community, it is not surprising that [the] diverse array of support from all sectors of the community is viewed as "unimpressive." Clearview is obviously more impressed with its role as a master antenna than with responding to the voice of the community.[18]

Almost in synchrony, Warner Cable withdrew from various agreements with local community investors, leaving minority investors' franches, many highly innovative, void for lack of capital.

One industry analyst, suggesting that subscription television could provide the capital necessary for development of two-way services and hardware, warned that "if pay TV is implemented as a one-way service, the future of new two-way services may be postponed indefinitely."[19]

Walter Baer, in a discouraging "summary" position paper, believes that cable will continue to evolve commercially, with "considerable discussion of, but little movement toward, alternative paths of development.[20] Edwin Parker and Donald Dunn, in a much-cited 1972 article, predicted innumerable advantages for our society through the use of two-way "universally available" cable networks; yet what *has* come true of their predictions is the development of the expected information gap between rich and poor.[21]

It is becoming increasingly evident that there are tremendous potentialities in the cable medium that will never be realized if the medium remains in the hands of private interests, regulated by a complaisant Federal Communications Commission. The advantages to local government of a municipal channel will remain unknown, despite the pleadings of the Cable Television Information Center for more experimentation on the part of the cities.[22]

> Mountains of reports and articles have been written about the potentials of this new medium, with almost all reports predicting an enormous social value for these urban CATV systems. . . . But the social, political, and economic forces in our communities reveal a

> somewhat different reality. There exists a gap between the people knowledgeable about cable television, and those who live in the communities to be wired for cable service. It is not only an information gap, it is a power gap.[23]

The problem is indeed one of power. For years, the cities, on behalf of their citizens, have played a waiting game with the FCC and the cable industry. Meanwhile, powerful economic interests have been working in the financial and political realms to consolidate their own power as the "cable industry."

THE CORRUPTION OF POWER

There is a tendency among communication planners to place an overwhelming faith in the efficacy of advanced technologies, as if the technologies themselves can right the wrongs perpetrated upon the urban dweller by the forces of the market economy. LeDuc wishes cable to be integrated "into the existing electronic media system" in order to avoid its congealing into another mass entertainment medium.[24] Anne Branscomb wants cable to become a common carrier, with open entry to the admittedly antagonistic AT&T and conventional broadcasters.[25] These planners hold onto an unwarranted expectation that once the "warring" among communication technologies is ended, cable will be freed to solve the problems of the masses.

But it is not a "battle of technologies" that has scarred the American urban scene but the battles—and collusions—of major corporate entities, whose sense of responsibility ends with the last shareholder. Sydney Head has detailed how "increased identification of broadcasting with the country's major corporate power structures" has created an obsession among broadcasting executives with the concepts of "market size" and "growth" as parameters for the measurement of success in that industry.[26] Likewise, cable managers have sought to maximize profits for large institutional investors, ultimately at the expense of local community members, whose unique needs do not conform to the corporate strategy.

The concentration of ownership within the cable industry is already so great that it may be impossible for cities to deal with operators in any manner approaching parity, as LeDuc has claimed.[27] The ten largest cable operators control an inordinate share of the total number of subscribers in the country, while the largest operator, TelePrompter, controls systems servicing over 12 percent of the total subscriber population. Most other systems are controlled by nominally competing media, such as television and radio stations and newspapers. The current recessionary trend in the economy has probably accelerated the trend toward consolidation of cable operations.

There is a point at which cable operators simply "outgrow" personal loyalties to local communities and their needs, and the results may be very injurious to the role of cable in the community, as has been the general case with American business enterprise:

> As it has never been demonstrated that the giant, expansionist state benefits the ordinary people who are regularly called on to sacrifice themselves in its behalf, so, too, though in a more subtle fashion, it appears that business firms have tremendous expansionist drives quite unrelated to the welfare of society in general.[28]

Michael Botein has directed attention to the problem of accelerating cross-ownership of cable systems by competing communications media, "inject[ing] capital but no new talent or attitudes into a new medium," and utterly stifling the development of diverse programming.[29]

The corporate hegemony penetrates into the chambers of the Federal Communications Commission. When the cities of Anaheim, California, and Arlington, Virginia, located cable operators willing to supply services exceeding FCC minimum standards, they found the FCC unwilling to issue certificates-of-compliance until the franchise agreements were suitably downgraded.[30] The situation has touched off a conflict between the FCC and a growing number of disgruntled cities, which may escalate into all-out legal warfare in years to come.

But for citizens to achieve a position of power will require more than mere reform of current regulatory practice. As Herbert Dordick has observed, franchise renewals and other regulatory leverage used to encourage private operators to wire "those areas of the city where the city believes there will be the most utilization and benefit" have been unsuccessful in every case.[31] It is a chicken-and-egg situation, which is not susceptible to incrementalist approaches advocated by Baer[32] and other reformists. In Dordick's words: "Clearly, a prerequisite for more widespread action awaits [the cities'] greater awareness of the importance of improved cable communications. And this, in turn, requires a greater appreciation of the role of communications in city government."[33]

How is this awareness to be achieved? FCC Commissioner Benjamin Hooks, admonishing operators, has stated that "public service is rarely 'cost-effective'."[34] But, Douglass Cater of the Aspen Institute complains, "the whole communications business is too often treated by its spokesmen as the last preserve of rugged individualism."[35] There is a pervasive ideology of capitalistic individualism, aided and abetted by liberal idealism, which requires debunking before any progress can be made toward public ownership—and utilization—of our communications resources.

Fortunately, there is a countervailing ideology, based on community cohesion and cooperation, which is as historically American as "rugged individualism." Murray Stedman, calling attention to the "very strong psychological attraction" of community control, suggests that "the movement for community control has a very profound basis . . . its roots go very deep. Beyond doubt, the drive for greater community control will continue to gain momentum."[36]

Commercial cable operators are incapable of meeting the communication needs of the cities—only ten percent of all Americans are able to obtain any cable services, and these services are only conventional retransmission or pay-movies for the most part. Informed observers have suggested that public ownership could be a truly liberating force, reducing the oppressive regulation of the FCC and increasing the diversity of programming and services available over the cable.[37] A city operating its own system could specify any feasible services for provision to its subscribers, without encountering obstructionism from operators and their regulatory allies at every step of the way.

Cable communications could provide a technological infrastructure supplying new channels of communication to accompany improvements in the urban social structure. It is this possible symbiosis between technological change and social change that can give hope for the urban future.

NOTES

1. California Council of Intergovernmental Relations, *Local Government Planning in California* (March 1972), p. 47.

2. Richard C. Leone and Robert S. Powell, Jr., "CATV Franchising in New Jersey." *Yale Review of Law and Social Action* 2, no. 3 (Spring 1972): 254.

3. *Cable Information Newsletter* 2, no. 7 (July 1973): 2.

4. "Philadelphia Group Goes after Basic FCC Cable Policy," *Broadcasting* (February 25, 1974): 53.

5. Don R. LeDuc, "Cable-TV Franchising and the Local Bar Association," *Wisconsin Bar Bulletin* 47, no. 6 (December 1974): 41.

6. Don R. LeDuc, *Cable Television and the FCC* (Philadelphia: Temple University Press, 1973), p. 207.

7. "Cable Presses for Tiered Regulation in Comments in Definition Proceeding," *Broadcasting* (November 10, 1975): 53.

8. Jerrold N. Oppenheim, "Cable TV and the Public Interest," *The Progressive* (March 1974): 47.

9. "More Local Programs with Better Quality Going Over the Cable," *Broadcasting* (September 9, 1974): 62.

10. "Cablecasting: A Myth or Reality?," *Rutgers Law Review* 26, no. 804 (Summer 1973): 37.

11. "FCC Rethinking Its Mandatory CATV Local Origination Rule," *Daily Variety*, February 13, 1974, p. 4.

12. *Report and Order*, Docket no. 19988, FCC 74-1279, 49 FCC 2d 1090 (1974), p. 1109.

13. "Whitehead Urges DJ 'Temper' Anti-Trust Activities Vs. CATV," *Daily Variety*, February 13, 1974, p. 4.

14. "Poverty Is Major Issue in Comments on '77 Deadline," *Broadcasting* (March 24, 1975): 38–39.

15. "In Cable Franchising, The Cards Are Being Played Closer to the Vest," *Broadcasting* (February 4, 1974): 48.

16. "No Raise, No Service," *Broadcasting* (October 14, 1974): 42.

17. "In Oakland, a Cable-TV System Fails to Live Up to Promises, Irking City and Minority Groups," *Wall Street Journal*, September 25, 1974, p. 30.

18. "A Review of the Arguments; Warner Backs Away," *Cable Information Newsletter* 3, no. 3 (March 1974): 2-3.

19. W. E. Himsworth, *Cable Television–An Industry Analysis* (New York: Salomon Brothers, February 29, 1972), p. 13.

20. Walter S. Baer, *Cable Television in the United States–Revolution or Evolution?* (Santa Monica, Calif.: Rand Corporation, 1974), p. 9.

21. Edwin B. Parker and Donald A. Dunn, "Information Technology: Its Social Potential." *Science* 176 (June 30, 1972): 1396.

22. Cable Television Information Center, *Local Government Uses of Cable Television* (Washington, D.C.: The Urban Institute, 1974), pp. 3–4.

23. Community Video Center, *A Proposal for an Urban Video Network for the Benefit of the Washington, D.C. Community* (Washington, D.C.: Community Video Center, 1971), pp. 1–2.

24. Don R. LeDuc, *Cable Television and the FCC* (Philadelphia: Temple University Press, 1973), pp. 205-7.

25. Anne W. Branscomb, "The Cable Fable: Will It Come True?," *Journal of Communication* 25, no. 1 (Winter 1975): 44-56.

26. Sydney Head, *Broadcasting in America* (New York: Houghton Mifflin Company, 1972), p. 319.

27. LeDuc, *Cable Television and the FCC*, p. 214.

28. Robert A. Dahl, "Governing the Giant Corporation," in *Corporate Power in America*, ed. Ralph Nader and Mark Green (New York: Grossman, 1973), p. 23.

29. Michael Botein, "Access to Cable Television," *Cornell Law Review* 57 (February 1972): 433–34.

30. "The Anaheim-Arlington Connection," *Broadcasting* (August 5, 1974): 48-49.

31. Herbert S. Dordick et al., *A Study of Local Government Uses of Cable Television and Integrated Communication Systems* (Los Angeles: Center for Communication Policy Research, University of Southern California, 1974), p. 58.

32. Walter S. Baer, *Cable Television in the United States–Revolution or Evolution?* (Santa Monica, Calif.: Rand Corporation, February 1974).

33. Dordick et al., *Local Government Uses of Public Television*, p. 58.

34. Federal Communications Commission, Docket no. 20363, FCC 75-821 (released July 22, 1975), concurring opinion, p. 6.

35. William L. Rivers and Michael J. Nyhan, eds., *Aspen Notebook on Government and the Media* (New York: Praeger, 1973), p. 11.

36. Murray S. Stedman, Jr., *Urban Politics* (Cambridge, Mass.: Winthrop Publishers, 1972), p. 310.

37. "Towards Community Ownership of Cable Television," *Yale Law Journal* 83, no. 8 (July 1974): 1729.

CHAPTER

6

THE CITIZENS TAKE CONTROL

SYSTEM PLANNING: "AMATEURS" VERSUS "PROFESSIONALS"

Democratic citizen involvement in the management of a public cable operation, in pursuit of progressive aims, would be as important as the achievement of public ownership itself. Overprofessionalized municipal management has too long served as a buffer between exploitive interests and the exploited masses, ameliorating tense confrontations between the oppressors and the oppressed through the ritual of "public participation." Today, this management strives to make the city "more efficient," eliminating jobs and services, and in so doing, greasing the gears of a social invention that is becoming every more difficult to keep in good repair.

If one issue preoccupies the minds of officials of local government, it is not the all-important management of information exchange in the city but the market-directed need to provide transportation facilities for the movement of workers and goods. In the period 1966-75, according to Alan Rabinowitz, transportation received "by far the largest slice of the (governmental) capital outlay pie . . . at $151.7 million, with the local units–states and cities–picking up $141.1 million of the tab, or a whopping 93%!"[1] This represents nearly half of the $327.8 million spent for local facilities of all kinds by local governments during the same period.

What issues has the National League of Cities considered to be "relevant"? A list used by the League for surveying citizen attitudes is both revealing–and disappointing:[2]

Public schools	Trash and garbage collection
Police night patrols	Inspecting, citing owners of

Giving tickets and towing illegally parked cars
Street lighting
Cleaning and repairing streets
Medical care for the indigent
Cleaning up parks for children
Welfare for children
run-down houses
Helping kids on drugs
Building low-cost housing
Controlling air pollution
Public transportation
Freeways
Recreation centers

It is as if cities are expected to do no more than provide remedial services necessary to keeping urban life tolerable, if not rewarding.

If the cities are to move beyond mere remediation or urban problems to actual rehabilitation of their environments, the task will most likely fall to that breed of professional managers known generically as *planners*. Mel Scott, official chronicler of the American Institute of Planners (AIP), believes that planners, impelled by the decay of the urban habitat, are beginning to realize that

> all forms of planning—economic, social, and environmental—must be linked, difficult as it may be to develop a truly encompassing planning process. The rediscovery of this basic principle sustains a new conviction: *social change not only can be planned but also must be* (emphasis added).[3]

Planner Mark Heyman, however, has observed that the official line advanced by the AIP and its members is more rhetorical than practical:

> Influenced by the social movement of the 1960s, the American Institute of Planners deleted "land use" from its definition of the profession as it appears on the cover of the *Journal*. But the practitioner—urged by academics, social scientists, and social activists to be concerned "with people instead of buildings and land"—has been shown few, if any, methods to replace land-use planning. Consequently, current planning practice is a long way from what planning schools say planning is, or should be.[4]

If planners are to be entrusted with the development of advanced forms of urban communications systems, then the picture painted thus far is certainly distressing: cable communications, if not first given over to the machinations of pragmatic bureaucrats, will be thrust into the hands of a category of professionals unequipped to function outside a very narrow domain.

But is the situation really so desperate? Have not planners frequently been identified with the "public interest," battling developers and commercial interests bent on destroying the most humane characteristics of the urban environment? Has not the strong moral commitment of planners resulted in a new school of planning, "advocacy planning," devoted to work on behalf of benighted citizens?[5]

In reality, planners too often have pretended to a rationality and objectivity that has disguised their own values and purposes—values and purposes often not in sympathy with their moral pronouncements. As James Berson has warned, "plans as well as goals are value-laden, and the plans set forth by whatever agency reflect the values of those who participate in creating them."[6]

What, then, are the values of this select group, the planners, who might, under prevailing circumstances, be entrusted with the development of publicly owned cable resources?

One major influence on urban planners has been the work of Russell Ackoff, a pioneer in quantitative methods of urban research and a strong proponent of "system theory." Ackoff recommended that planners use the city as a laboratory for evaluating theoretical models, suggesting that "social experimentation must become a common tool of the planner."[7] A problem, however, was that planners did not have the right attitude: "I see little evidence of [the planner] trying hard to experiment," even though the planner had little to fear, since such experimentation only "requires acceptance by the public that will be affected by the required manipulation."[8]

Ackoff's theories found wide circulation among the planners in the federal bureaucracies of the Kennedy and Johnson administrations: these planners, recruited heavily from industries and academic institutions attuned to hierarchical modes of management, were not averse to putting Ackoff's prescriptions for "required manipulation" into practice. The "systems" orientation of these planners, at both the local and the federal levels, survived the trauma of the Nixon-Ford years and pervades planning departments today. What opposition there has been to quantitative, systems-based planning, with its strategies for "intervention," has been of a mostly conservative variety. Peter Szanton, for example, does not question the basic tenets of elitist systems-based planning (conducted invariably by extrasystem planners); rather, he only demands that the theoreticians "get their hands dirty" with the politics of decision making more often than they presently do.[9]

Recently, however, a truly critical school has grown up within the planning profession, and it is this school which gives hope that at least some planners may be up to the challenge of planning for urban communications systems. The members of this school do not deride conventional planning, either systematic or pragmatic, for any lack of grit or efficiency; instead, they attack its basic premise that "social change must be planned." The adherents of this critical school of planning do not reserve the prerogatives of planning to an exclusive clique of elite professionals: they seek to make their clients—the "average" citizens—share at least a part of the responsibility for the decisions that affect their environment and their lives.

Charles Jencks and Nathan Silver, in the avant-garde of this movement, assail conventional planners as advocates of a brutal status quo:

> . . . the means of system-making, categorizing and organizing have emerged in a newly obstructive role. These means of more "efficient," less responsive control are the most typical and disturbing of our time. To an unprecedented extent we now try to master-plan and control changing aspects of culture and society. What's wrong with the world has become not its disorder, but its repressive order.[10]

Jencks and Silver identify three types of planning, which they call *master planning, piecemeal planning,* and *contingency planning*; of the three, only the last escapes susceptibility to "crude excesses of unresponsiveness."[11] Championing a new social-action philosophy, *adhocism*, Jencks and Silver claim that contingency planning techniques can prevent conscious and unconscious repressive acts on the part of the planner:

> Behind the idea of contingency planning is a simple willingness to take feedback into account. Feedback (especially from the community) makes it possible to develop new rules that recognize earlier successes, and to improve already existing but less successful things, and to verify a well-tested model for the future.[12]

The contingency planners, practicing adhocism, have turned systems planning (as traditionally employed) on its head: the systems paradigm is not rejected, but the objective of systems planning from an adhocist point of view is not to test grand theories but to enable urban people to improve the conditions of their own lives–an orientation that can accommodate the need for broadbased planning for cable communications.

Alexander Tzonis has moved even further toward eliminating the elitist nature of most urban planning–elitist, perhaps, because of the defensive posture planners are so often forced to acopt–in issuing a call for planners to abandon the formal processes of design, which characterize planning's role as preserver of cultural "ideals." He traces the history of planning methodology, fulfilling its own propecies of social oppression, to the 1700s, when mercantile capitalism was transforming the Western world and calling forth the new science of "'Mechanicks' . . . a means to advance the concentration of society . . . where power has to be concentrated, secured as well as multiplied."[13] The rise of "mechanicks"–the forerunner of architecture–in turn strengthened the concept of "structural efficiency," which provided "a justification not only for those who benefited power but also for the oppressed, allowing for architects the image that they operated on principles which benefit the totality of society.[14] The resulting implication that architecture–and now urban planning–is nonideological and that architects and planners are the most capable instruments for bringing the urban environment "into line" finds its roots in the philosophies of one of the most exploitive periods of modern human history.

Tzonis's observations on the endless chain upon which conventional, "rational" planning and design are intertwined deserve a more lengthy explication, for they cast great doubt upon the ability of planners to successfully design and implement such complex and important urban systems as cable communications. But one passage will suggest the essence of Tzonis's critique:

> The rationalization of architecture was presented as an urgent obligation, an essential necessity for survival in a critical technological situation. The application of rational methods to other activities of man, to production, to transportation, brought about needs that never existed before. These needs presented great difficulties, they became problems, environmental problems, problems of man's habitat. Man was seen to be oppressed by conditions of extreme deprivation. His very survival was threatened by his environmental conditions, the conditions of his habitat. The designer [or planner] was expected to face this "problem" and provide "solutions" in order to avert a situation of "major catastrophe." . . .
>
> The rationalization of design failed to resolve the crisis. Not only did it fail to bring a non-oppressive environment, but it assisted in furthering oppression in the man-made environment because it considered the limitations of designers [or planners] results of their problem solving methods, inadequate for producing successful products, where there was nothing to be solved. *The non-oppressive environment does not depend on the quality or quantity of the products [or plans] which populate it but on the relationship of its inhabitants* (emphasis added).[15]

Tzonis's conclusion goes against the traditional wisdom of urban planning (and communication planning after it) that the mere manipulation of physical objects—the use of technology—can adequately address the problems of the urban environment. If one accepts Tzonis's argument, then the success of a cable communications system within a larger human environment cannot be predicated on the technological competence of the system but must be dependent on the equity of relations among the human beings who are, in one way or another, tied to the system.

The participants in the Delos Seven conference in 1969 urged planners contemplating change strategies for the city to take a second look at their "laboratory," seeing it in its fuller dimensions:

> We can no longer use the word "city" without thinking twice because we all have an image of the city as a built-up area, and we forget that the ordinary man and his family no longer live within a city but within a system of life processes which extends well beyond the city, which it only touches at certain points.[16]

Maurice Ash has expanded on this warning, noting that "the approach to urban form by way of the 'built environment' ends up in an architectural strait jacket upon society, representing a stereotyped and unreal idea of contemporary life."[17] Ash's point is a simple one: planning, in its preoccupation with spatial phenomena, has become an overly complicated and manipulative process, excluding from its practice the very people who should be a part of it–the citizens! Ash writes:

> If planning is about our neighbors, it should after all be possible for each to understand it. And, until of late, we have supposed we did understand it: that we knew what a city was, or a town. The misunderstandings between us, however, are now proliferating. In our unending game of words, situations are developing around us in which, try as we may to keep pace with one another, meanings elude us; meantime, others among us, to try to preserve the meaning of "city," make gigantic efforts to make the clock stand still.[18]

How, then, are citizens–and especially the effectively powerless urban masses–to be incorporated in the management of a technology such as an urban cable communications system when the planners themselves are unwilling to commit resources to their study and understanding? The "game of words" of the professional planners must eventually come to an end, but correct social action will have to be taken long before it does.

If the public cable enterprise is to truly serve the public, which underwrites its existence, citizens will have to be involved in its development at every step of the way. The planners, unfortunately, cannot be counted on to provide direction; nor should the technological package simply be tacked on to the present dysfunctional public bureaucracy.

A widespread appreciation of the potential of telecommunications technology for upgrading the urban environment–as well as an awareness of the possible dangers inherent in the technology–is a necessary first condition for effective planning in this area. Planners could, of course, play an important part in generating this awareness and act as catalysts of social change around the issues of urban communications and related technologies. But their success will depend on their enlistment of equally foresightful and progressive allies among the "amateurs" of city affairs–the citizens who daily "live" the meaning of the city. Not all planners will want to take the chance of adventuring into this region of uncertainty, previously dominated by engineers and the corporate managers whom they serve; those who do, however, will have to abandon their former elitism and their spatial biases in order to build strong coalitions able to confront corporate and governmental foes in the economic and political arenas.

The challenge is one befitting the near end of a millinium, for "amateurs" and "professionals" alike.

REORDERING THE FLOW OF COMMUNICATIONS

Paul Baran early grasped the full potential of cable communications for totally reshaping the city in America: he speculated that the substitution of communications for transportation "could mean the elimination of much which the city of today offers, indeed of its very raison d'etre . . . ," access to the variety of experience possible in the dense urban core.[19] He worried, however, about the "communication-dispossessed," and their inability to survive in an increasingly impoverished and empty urban shell.

Baran examined the role telecommunications could play in interceding between the citizen and his government—and the discord that could result if the need for this intercession was not satisfied:

> The flood of news flooding downward and the felt lack of an upward channel of reply can be expected to worsen in the near term. We need an information feedback channel to assure the citizen that his specific message (not the group response) has, in fact, been received by the power structure. Failure to supply such a channel would appear to guarantee alienation from the political process. No one has yet seriously considered providing the ombudsman junction electrically. Yet in the large cities it may be the only realistic course open.[20]

Neither the point-to-point telephone system nor the one-way broadcasting medium can build a collective identity sufficient to triumph over personal isolation endemic to the modern city. Point-to-point telephones, while highly interactive, can support only fragmented and isolated instances of communication: the notorious "telephone tree" of political campaigns is a clear example of the inefficiencies of mass action—necessarily hierarchic, authoritarian mass action—conducted via the inflexible telephone network.

One-way broadcasting is even more deleterious: not only does it discourage collective identity among scattered millions of viewers by programming only fantasy experience but, according to Raskin, it promotes the longevity of an ideology favoring the exploitive interests that sponsor its propagandistic activities:

> The TV medium is an example of how isolated and separated from one another people are. At any one moment millions may be listening to a specific program, but reinforced by the nature of television itself no one would think there is any control he could exercise with others. . . . TV becomes the pyramidal instrument to give the appearance of understanding participation and sensual participation, so that the dissonance of one's life, which could be activated through personal recognition of one's feelings and related to the dissonance which other people feel, is destroyed.[21]

There is no assurance that cable will develop into a medium any more capable of reducing dissonance by encouraging self-awareness and collective action. As Richard Hooper has lamented, in a criticism of too-rapid franchising by some municipal authorities, "if the only two-wayness (of cable systems) is response buttons of the yes/no variety, then broadcasters who already use phone-in and write-in response systems will remain more than competitive."[22] Unfortunately, as Hooper notes, only a literal handful of experimental systems have even that limited capability.

There is, moreover, a distinct possibility that the introduction of cable communications under present forms of management (beholden to exploitive economic interests) will exacerbate the social inequities of urban life by making action-stimulating communication even more restricted and exclusive. Grady Clay, a journalist and observer of the "urban beat," has charged that

> corruption feeds on monopoly—on the ability of small groups to occupy and manage essential vantage points in the flow of information—data centers, official agencies, large corporate research facilities. Such vantages ought to be exposed to public surveillance and review. It is the public itself, rather than a managerial elite, which ought to have fullest access to the vantages I have described.[23]

A cable bureaucracy once rooted in place could become totally self-interested and unresponsive to citizen needs. It would be far easier to prevent the creation of such an exclusive "vantage point" than to attempt to extract it from the urban milieu once it was in place.

Benjamin Singer, after studying "citizen feedback" in Canadian society, concluded that "in a fully developed democracy with the maximum of technological channels of communication, the effective use . . . of feedback becomes choked through the sheer weight of the difficulty of trying to make oneself heard."[24] Once resignation has set in, "the area of unchallenged administrative decision making expands even under the mantle of a 'participatory democracy'." Singer emphatically declared that

> *actual* participation becomes the critical issue; and actual participation is not synonymous with *potential* access to channels, for access per se tells us nothing about the difficulties of the process itself for individuals differentially located in the social system.[25]

Access of two distinct, and equally essential, kinds must be incorporated in any anticipated municipal cable utility—access to the medium itself and access to the direction and management of the operations of the medium. In actual planning for such a utility, it would be extremely important to take precautions to avoid the historically demonstrated tendency of communications managers to seek the reduction of public participation in the operations of their electronic domains.[26] Citizens must assert their claims at the onset of planning, for otherwise management will become an increasingly powerful political force, able to restrict access and secure a position as "gatekeeper" of vital information.

John McHale has warned that "the new information environment" being created today by the new media of communication could "create a new and powerful 'property class,' whose property is in their heads, i.e., those possessing uniquely the specific skills required for access to, and manipulation of, vital knowledge and information."[27] There can be no doubt, in light of the illusiveness of access without ownership and participation by an informed public, that planning for a cable communications system in the city must include involvement of the urban masses.

Unfortunately, this involvement will probably have to wait upon the initiative of the citizens. It has already been shown how unaware most American cities are of communication phenomena occurring in their midsts.

Even in progressive cities, "right now, cable has low priority."[28] Cities surveyed by the Annenberg School of Communications at the University of Southern California ranked the possible uses of urban telecommunications systems in the following order: (1) citizen information, (2) civic education, (3) issue formulation and discussion, and (4) citizen participation.[29] It is significant that these municipal managements view the cable utility primarily as a tool of government to inform and educate citizens, and only secondarily as a device for stimulating public debate and involving citizens in civic affairs. The hierarchic forms of local government, which are the rule today, will have to give way to a more "adhocist" and democratic form of government if cable is to be allowed to augment participatory management. This is hardly a trivial challenge for those trying to help the cities implement communication technology.

Yet, this effort will have to be accomplished if cable is to come to the city for more progressive reasons that advocated by those seeking a "safer" entree for the medium:

> Government officials will be most interested in long term attitudinal changes reflected in voting behavior. . . . It will be difficult to determine just what the contribution to the mayor's re-election a specific information campaign or improvement in information delivery practices has actually been. But such effects are crucial to local government's acceptance of communications *as a significant service for their citizens* (emphasis added).[30]

If cable is to mean more than additional votes stuffed in a ballot box every fourth year, citizens will have to see that the government which owns and operates it is absolutely accountable and as democratic as possible. It is idealistic to imagine that politics and the media of communication can be isolated one from the other; citizens, in anticipation of the aid they will receive from a citywide communications infrastructure designed to encourage collective actions, must begin the transformation of local government before system design actually is initiated—and make the politics of the city truly the politics of the people.

In light of the poor performance of municipal government structures in the past two decades in America, it would not be surprising to see at least some progressive city officials assisting in this major task.

WHAT SCALE FOR SYSTEM DEVELOPMENT?

As Edward Roth discovered from his vantage as a seasoned urban manager and consultant, full community involvement with due process for every citizen must precede any successful cable planning: "Unless a community supports a proposal for municipal ownership the entire matter is rendered moot."[31] Most cities would be hard pressed to fulfill this basic requirement; the hierarchical method of management, inherited from the corporate model, discourages citizen involvement in decision making. The "town hall" is not dead, but it is incapable of being resuscitated until a means for citizens-to-citizens communication is realized in a flexible urban telecommunications network. (This is a "bootstrapping" problem that, abetted by the NCTA and the federal authorities, will hamper efforts to launch a successful municipal system.)

Given the present urban environment, however, there is every indication that citizen pressure upon municipal managers will force the managers increasingly to discard their identification with the profiteering exploiters of the city—the banks, the developers, the real estate and utility owners—in favor of a catalytic role seeking to encourage urban collective action. George Washnis has described the present tendency toward municipal decentralization:

> Everything seems closer now . . . even in the middle- and higher-income areas, issucs of resource distribution, crime, and community stability have aroused residents so that they too want to be part of the decision-making process. It is more and more up to the elective and administrative leadership to devise workable local systems of community interchange and shared power.[32]

But what are these communities—and which "elective and administrative leadership" should devise workable systems? Since it has already been established that the isolated community will not be able to create a cable system on its own, should the agent of the communities be the city, the county, the state, or some other special entity?

Foundation 70, a public interest group, has argued that the city can properly provide for the development of people-serving cable systems (in this case, franchised systems),* for "if systems are planned with careful attention to existing communities in the city, allowing for migratory trends and redevelopment plans . . . cable communications could be a tool of community cohesiveness."[33]

An excellent manual, which has never received adequate public attention, was prepared by the San Gabriel Valley (California) Public Cable Council specifically to answer the question, "What form for public ownership?"[34] A team of legal and economic consultants, directed by Louis Reiss of Pasadena City College, analyzed three alternative forms of public ownership—municipal ownership, municipally sponsored public corporation ownership, and ownership by a regional special authority. (An example of each kind of development is described in the Appendix.) After a long and intensive study of the relative merits of each form of ownership, the Council came to the following conclusion:

> A Special Public Authority may be the best means of providing cable television services to contiguous geographical areas encompassing more than local government jurisdiction. Such regional service may be desirable in terms of economy (one head-end and studio facility; one set of management and administration personnel; larger market area and revenue base to justify more sophisticated technology and programming) and in terms of uniting with one communications link interdependent areas having common interests. If a cable television system is to be limited to a single jurisdiction, then it should be under direct control of general purpose local government.[35]

Though Reiss's team obviously favored the economies of a larger geographically distributed system, it did not reach a definitive conclusion as to the appropriate jurisdiction for cable management—joint-powers agency, individual cities, or some other entity.

There are theoretical positions that are critical of the special or joint-powers authority as an easy solution to the problem of local inertia. For instance, special authorities and local autonomy are often held to be antithetical.

---

*Foundation 70 would probably not be so optimistic today regarding the power cities can manifest in dictating franchise terms: no city has ever been able to satisfactorily instruct a franchisee as to where cable should be extended or what access should be provided. In fact, the power of the cities in dealing with private operators has been severely curtailed by the FCC in recent years. The situation today is more akin to that which Foundation 70 warned against when it said: "If a city, whether by franchising a number of systems or one system with many subheadends [local program-originating facilities], permits an arbitrary pattern of development, many 'natural' communities may be cut and fragmented in communications terms just as many have already been fragmented by highways" (Foundation 70, "A Cable Is a Very Big Wire," *Yale Review of Law and Social Action* 2, no. 3 [Spring 1972]: 205).

A Canadian critic, Walter Stewart, has noted that while "city planners . . . are concerned to keep the downtown areas as varied, cosmopolitan and alive as they are today . . . regional planners . . . tend to regard the city as, logically, a place to work for suburbanites: downtown is the factory, outside the dormitory."[36] Questioning "the trend to lump municipalities into regions," Stewart warns that "until power and cash can be made to follow responsibility, it is hard to see how any permanent solutions will be achieved."[37]

In Hartford, Connecticut, city planners looked toward "metropolitan government" as a way to get wealthier suburbs to "share their financial and intellectual resources with the central city, presumably for the common benefit of all."[38] Expectedly, reaction among the citizens of the middle-class suburbs was not positive.

Surprising, however, was the reaction from urban inhabitants, the ghetto dwellers, who perceived a threat posed to their collective power by metropolitan government:

> The supposed beneficiaries of metropolitan government—especially the Blacks and the Puerto Ricans in the North End of Hartford—began to reassess the demographic realities of metropolitan government. Under the existing system, they had a growing voice in the politics and government of the city, particularly in the public school system. Under the consolidation plan, the city population of 160,000 would have been swamped in the total regional population of half a million. By the end of the 1960-70 decade, leaders of substantial minority groups in Hartford were understandably extremely chilly toward metropolitan government. Their reservations on this subject were widely shared throughout the city.[39]

Like metropolitan government, metropolitan cable would undoubtedly stir passionate feelings among populations of the urban core and the suburban ring.

Given the procedural difficulties of establishing a regional authority, the city emerges as the governmental unit of choice—at least for large cities—for the development of local cable services.

Cities might secure assistance from the local county. Alan Campbell and Seymour Sacks examined nationwide patterns of metropolitan and nonmetropolitan governments; they determined that metropolitan counties can supply financial support far in excess of that now supplied to their municipal clients.[40]

With virtually undeveloped tax bases, despite levels of expenditure already in excess of those of most cities, the counties could act as nuclei in the formation of metropolitan cable systems—as capital investors rather than as managers. A modified "Lakewood Plan"—whereby the county provides services at cost to its "contract cities"—could be implemented in those states where cities and counties are empowered to enter into such agreements. Smaller cities, however, might wish to examine the special authority option more closely, especially if they share common aspirations and goals with resource-limited neighbors.

STEPS TOWARD PUBLIC OWNERSHIP

A city contemplating municipal ownership will first have to investigate the state laws of incorporation, which may or may not empower the city to operate a cable communications system. William Moore, secretary of Jacksonville, Florida (which already runs a municipal broadcasting station), believes the benefits of specific empowering legislation would be threefold:

> [it] would easily cure the problems of questioned authority. . . . [and] provide those legal benefits for a cable authority which it has provided for many . . . other independent agencies of governmental or entrepreneurial nature:
> (1) eliminate doubt of the agency power to function and to issue securities;
> (2) expressly detail the structure, purpose, financing and operation of the activity;
> (3) relate the agency to the City by sufficient political ties to assure service to the City, yet supply that measure of independence necessary to a professional and non-political purpose.[41]

Moore's caution is well-taken: the cities will have to reassure their citizens that their authority to operate cable communications systems is genuine.

Having obtained state sanctions allowing it to enter into financial contracts for the construction of the system, the city would have to perform necessary prefinancing engineering studies. These would be required to determine the actual cost of the system and a schedule for construction by the city. (Herbert Dordick has insisted that a large city could use existing conduits and, where they exist, municipal power poles, to carry cable to residences and businesses at a fraction of the costs incurred by private operators in pole rentals and undergrounding.[42])

Alternatively, a city could enter into a "mixed-ownership" agreement with a private operator, each partner sharing a portion of the costs of construction and the channels made available for programming and service transmission.[43] (If all signals were encoded in digital form, many more transmissions could be multiplexed onto available channels, increasing the carriage capacity of the system.) The city would act as agent for all noncommercial institutions and citizens; the private operator would offer conventional pay-TV and rebroadcast programming.

The city would next have to evaluate its ability to finance the public ownership of all or part of the system. If sufficiently endowed (either with internal reserves or revenues from a profitable utility), it could finance the system through an internal transfer of funds. The costs could be amortized, as loans, over an extended period of time. This procedure would preferably be utilized on a one-time-only basis, unless there was some special reason for subsidizing system operation: self-sufficiency, through subscriber revenues and signal-carriage tariffs, would be a desirable early goal.

Cities without capital reserves or expected surplus income would want to consider approaching the county or state government for initial support; should the Congress approve some variation of subsidization plan along the lines of the Transportation Act, federal aid might become available for an initial investment. However, with each ascending level of government approached for assistance, the city risks the deflection of system goals for nonlocal purposes.

If the city decided to offer securities—general or revenue municipal bonds—through a process of referendum or other means, it would certainly want to consult authorities in this area to be sure it was pursuing the sale of the items in the most advantageous manner possible. A bond broker and a bond attorney could do more than prevent unnecessary liabilities: they could also locate potential buyers and negotiate for the most satisfactory rates prior to actual marketing of the issues.[44]

An increasingly useful resource in the issuance of securities is the Municipal Bond Insurance Association (MBIA), which guarantees tax-exempt municipal bonds.[45] Obtaining MBIA coverage could lower interest rates on a municipal bond offering by as much as four percent, and make possible the sale of securities for municipalities with less-than-excellent credit ratings. For an enterprise as uncertain as a cable communications system and with the total offering conceivably exceeding many tens or even hundreds of millions of dollars, bond insurance might be a necessary prerequisite to any sale.

Smaller cities and cities with inferior credit ratings would have to begin early locating underwriters willing to guarantee their bonds prior to marketing; possibly, bond insurance is in order. These cities would also need to gather preissuance capital for financing the activities of managers and citizens engaged in the bonding effort. The issuance of minibonds at least supply sufficient preissuance capital to complete preoperational planning; with increasing familiarity, this mode of "inhouse" financing may prove adequate to the greater tasks of construction and operations still lying ahead. (See Chapter 7 for more on minibonds as a tool for utility financing.)

As the issuance of bonds was completed, and capital became available for construction, the citizens could assume increasing responsibility for management of the partly constructed cable system; this would prepare them for their ultimate responsibility—operation of the completed system. Presumably, this would call for the creation of neighborhood associations, various work-centered organizations, public groups organizing around issues or recreations, and other foci for collective action. It would be arrogant now to presume to tell citizens how to organize or what structures to adopt in their organization. The whole purpose of public ownership is to allow for determination, by an involved and controlling public, of the new "soft system" for cable communications–the management and service components–without pressure from oppressive elites or their technocratic hirelings.

ALTERNATIVES REGARDING THE PRIVATE OPERATOR

While some coexistence arrangement could be arranged between the public system and the private operators, the municipality will most likely prefer to absorb the private system into the public network. There are two reasons for this: the private operator will be unable to offer the same services as the public system, since the technology of the private systems of today (and the foreseeable future) is and will remain inferior to that of the public system; and since the public system will, in one way or another, be financed from the public treasury, service must be extended to all citizens desiring service, including those already "on" the private system.

This does not rule out the possibility of a single community within the municipality maintaining its own cable network, if that is the popular desire; however, the city should be offering sufficient channels and opportunities for community control to obviate the necessity for such a costly redundant activity on the part of a single community. One of the reasons for creating a public system would be to unify the city and reduce dysfunctional informational disparities among its various communities; the existence of alternate systems would only serve to defeat that purpose–and jeopardize the effort to restore participatory democracy to the city.

Several recent cases indicate that cities could legally and economically compete with private operators.

In Monroe County, Florida, county officials won a court decision allowing them to construct a system of publicly owned "translators"–amplifiers which rebroadcast a television signal–that could eventually be in direct competition with a cable system the country had franchised earlier.[46] In its successful defense against a suit brought by the cable operator, the county cited the nine-year history of the franchise, in which service had been extended only once beyond the initial construction area–leaving most county residents without cable service.

In Puerto Rico, Governor Rafael Hernández-Colon implemented plans to establish a nonprofit corporation, replacing the inefficient Puerto Rico Telephone Company formerly owned by International Telephone and Telegraph Corp. (ITT).[47] Government consultants estimate the new company, run as a public trust, will save customers over 37.5 cents of each dollar spent by a comparable private company (AT&T) as taxes, interest on loans, and dividends to stockholders. (Mainland telephone companies are reportedly worried over the precedent-setting action of the American territory.)

In the cable field, a private operator unsuccessfully brought suit against the city of Frankfort, Kentucky, charging unfair competition: the city owns municipal power utilities, which grant the local public (nonprofit) cable company free pole rights (see Appendix). The city, in response to complaints from citizens that service by the private operator was inferior, had allowed the public cable company to overbuild the private system in a portion of the city. The private operator lost in both state and federal courts.[48] Ironically, the city granted the private operator the same free pole rights, despite the expiration of the contract for their use.

There is yet a last obstacle to be overcome by public systems attempting to incorporate privately served communities into their domain, the possible unwillingness of an industry-oriented FCC to grant certificates-of-compliance to municipal systems competing with already-certificated private systems. This situation remains hypothetical for the present, though it eventually will be put to the test. As Steven Rivkin has commented, "It would seem premature to conclude that whole fields of regulatory concern have been definitely preempted from local franchising authorities. . . . the launching of independent initiatives—possibly coordinated with other local franchising authorities—would ensure that local desires are vigorously pressed."[49] Rivkin's legal opinions on franchising are applicable to city options for direct operation: the FCC's jurisdiction over local cable matters was sustained by only one vote in the Supreme Court (in the *Midwest Video* case) and could be challenged at any time through Congressional action to modify the FCC's authority.

Thomas Brom and Edward Kirshner have noted that franchises patterned after the League of California Cities' model franchise contain a clause enabling the city to require "additional or greater standards of construction, operation, maintenance or otherwise"—a device by which a city could force an operator to default on a franchise without recourse.[50] Brom and Kirshner intend that no holds be barred: "The struggle for community ownership will not be an easy one. But cities have a lot at stake."[51]

Private competition, however, is not the main threat to public enterprise. The threat is apathy and carelessness on the part of the city and its citizens, which could damage the planning and financing of the projected system and preevent its ever coming into existence. Despite the skepticism of conservative lenders, cable systems have been, and will continue to be, planned and financed by cities willing to take the time and effort to ensure sturdy foundations for cable communications systems owned and operated on behalf of the people of the cities.

NOTES

1. Alan Rabinowitz, *Municipal Bond Finance and Administration* (New York: Wiley-Interscience, 1969), p. 181.

2. Floyd J. Fowler, Jr., *Citizen Attitudes Toward Local Government, Services, and Taxes* (Cambridge, Mass.: Ballinger, 1974), p. 38.

3. Mel Scott, *American City Planning Since 1890* (Berkeley: University of California Press, 1969), p. 642.

4. Mark Heyman, "What People Do in the City," *Planning* 1, no. 7 (August 1975): 26.

5. Anthony James Catanese, *Planners and Local Politics: Impossible Dreams* (Beverly Hills: Sage, 1974).

6. James H. Berson, "The Human Values of City Planning," *Journal of Environmental Systems* 1, no. 3 (September 1974), p. 283.

7. Russell L. Ackoff, "Toward Quantitative Evaluation of Urban Services," in *Public Expenditures Decisions in the Urban Community*, ed. Howard Schaller (Washington, D.C.: Resources for the Future, 1963), p. 116.

8. Ibid.

9. Peter L. Szanton, "Systems Problems in the City," in Rand Corporation, *Rand 25th Anniversary Volume* (Santa Monica, Calif.: Rand Corporation, 1974), p. 95.

10. Charles Jencks and Nathan Silver, *Adhocism* (Garden City, N.Y.: Doubleday, 1972), p. 180.

11. Ibid., p. 182.

12. Ibid.

13. Alexander Tzonis, *Towards a Non-Oppressive Environment* (Boston: i press, 1972), p. 74.

14. Ibid.

15. Ibid, pp. 93, 99.

16. Delos Seven, "The Scale of Settlements and the Quality of Life," in *Human Identity in the Urban Environment*, ed. Gwen Bell and Jacqueline Tyrwhitt (Harmondsworth, U.K.: Penguin, 1972), pp. 240–41.

17. Maurice Ash, *Regions of Tomorrow: Towards the Open City* (London: Evelyn, Adams & Mackay, 1969), p. 34.

18. Ibid., p. 10.

19. Paul Baran, "On the Impact of the New Communications Media upon Social Values." *Law and Contemporary Problems* 34, no. 2 (Spring 1969): 251.

20. Ibid.

21. Marcus G. Raskin, "The Dream Colony," in *Television Today*, ed. Richard L. Stavins (Washington, D.C.: Communication Service Corporation, 1969), p. 10.

22. Richard Hooper, "Yet More Spaghetti," *Urban Review* 6, no. 7 (November-December 1972): 6.

23. Grady Clay, *Close-Up: How to Read the American City* (New York: Praeger, 1973), pp. 179–80.

24. Benjamin D. Singer, *Feedback and Society* (Lexington, Mass.: D. C. Heath & Co., 1973), p. 5.

25. Ibid.

26. Stanley Scott, "Lessons from the History of Broadcasting," *Science* 178 (December 22, 1972): 1264–65.

27. John McHale, "The Changing Information Environment: A Selective Topography," *Ekistics* 211 (June 1973).

28. Herbert S. Dordick et al., *A Study of Local Government Uses of Cable Television and Integrated Communication Systems* (Los Angeles: Center for Communication Policy Research, University of Southern California, 1975), p. 28.

29. Ibid., pp. 24-25.

30. Ibid., p. 52.

31. Edward J. Roth, "Should City Hall Own the Cable TV System?," *Nation's Cities* 11, no. 6 (June 1973): 64.

32. George J. Washnis, *Municipal Decentralization and Neighborhood Resources* (New York: Praeger, 1972), p. 7.

33. Foundation 70, "A Cable Is a Very Big Wire." *Yale Review of Law and Social Action* 2, no. 3 (Spring 1972): 205.

34. San Gabriel Valley Cable Council et al., *Municipal Decision-Making Factors Relative to Cable Television Ownership* (Pasadena, Calif.: Pasadena Urban Coalition, 1973.

35. Ibid., p. 32.

36. Walter Stewart, "Why Cities Don't Work," *MacLean's* 87, no. 1 (January 1974): 64.

37. Ibid., p. 66.

38. Murray S. Stedman, Jr., *Urban Politics* (Cambridge, Mass.: Winthrop Publishers, 1972), p. 301.

39. Ibid.

40. Alan K. Campbell and Seymour Sacks, *Metropolitan America* (New York: The Free Press, 1967), p. 107.

41. William D. Moore, "Influences of State and Local Law," in *Cable Report for Jacksonville, Florida*, ed. Cable Television Information Center (Jacksonville, Fla.: Cable Television Information Center), addendum.

42. Herbert S. Dordick, of Annenberg School of Communications, University of Southern California, private communication, November 3, 1975.

43. Telecommunications Management Corporation, *Cable Communications for the City of Portland and Multnomah County* (Los Angeles: Telecommunications Management Corporation, 1973).

44. Gordon L. Calvert, ed., *Fundamentals of Municipal Bonds* (New York: Securities Industries Association, November 1973).

45. "Market Briefs: The Municipal Bond Insurance Assn.," *Business Week* (February 16, 1976): 72.

46. "Fla. Cable Fights with County Over Proposal for Competing Translators," *Broadcasting* (March 18, 1975): 91.

47. "Puerto Rico Disconnects ITT," *Business Week* (February 16, 1074): 52.

48. Robert L. Steiner, *Visions of Cablevision* (Cincinnati: Stephen H. Wilder Foundation, 1972), p. 35.

49. Steven R. Rivkin, *Cable Television: A Guide to Federal Regulations* (Santa Monica, Calif.: Rand Corporation, 1973), no. R-1138-NSF, pp. 77-78.

50. Thomas Brom and Edward Kirshner, "Cable TV: Money to the People," *Working Papers for a New Society* 1, no. 4 (Winter 1974): 42.

51. Ibid.

CHAPTER

7

THE ECONOMICS OF PUBLIC CABLE

MEETING UNMET NEEDS

The social imperative to invest public monies in an innovative technology, and to make the necessary and concomitant commitment to the institutionalization of popular participatory democracy in the city, can only come from a clear understanding of the inherent inability of private enterprise to provide a people-oriented cable utility—and a recognition of the costs of "information poverty" for the urban masses.

The misrepresentations of cable entrepreneurs in the early years of cable development have succeeded in casting public doubt and suspicion upon its potential, now labeled *blue sky*. A former officer of the National Cable Television Association explained the apparent reneging of the cable industry upon service promises made in the frantic franchising period of the 1960s by suggesting that "the cities have demanded too much" and that the cable industry has had to encourage false expectations "because it's all a function of free enterprise."[1] The notion that public services would be obtained from the private cable industry was a deceit used to secure lucrative franchises (to be traded away to the highest bidder).

TelePrompter Corporation, largest multiple-system operator and self-proclaimed leader of the cable industry, was the subject, in 1972, of an investigation by the Securities and Exchange Commission regarding allegations of false reporting to stockholders; as part of a corporate shake-up designed to "trim off the excess fat," Chairman Jack Kent Cooke slashed the MSO's budget, eliminating most support for local programming.[2] Other operators, coordinating their actions through the auspices of the NCTA, moved to eliminate their local services, including public access and local origination. By the end of 1974, the FCC, under pressure from the cable industry, had relieved the operators of local-origination programming obligations, substituting in their place an arbitrary and wholly inadequate public access requirement.[3]

Also beginning in 1974, most cable operators curtailed expansion of their physical plants, alleging an inadequacy of capital for investment. At the same time, they launched a nationwide campaign for higher subscription rates, less local regulation, and a chance to "experiment" with pay-cable. It seems unlikely that any amount of available capital could have persuaded the operators to expand their public service offerings.

Gustav Hauser, architect of Warner Communications' cutbacks and consolidation, spoke for the entire industry when he placed the blame for this state of affairs where it never belonged—on the public sector:

> Lured by the promise of big returns, cities demanded huge franchise fees—and got them. Rand, Brookings, Sloan, and other non-profit groups urged the cities to demand two-way capability, 48-channel systems, and other requirements for the future. The result is that systems were overbuilt at great expense and the facilities are not being used. Now we're in the position of having to explain that this industry can realize some profits, that some cities are buildable and profitable and some are not, and that cable-TV of tomorrow—with shop-at-home facilities, facsimile newspaper transmission, and so on—cannot be installed today.[4]

Hauser's misrepresentations of history—no two-way, 48-channel systems were ever specified, let alone delivered—have become the basic tenets of the cable industry's philosophy: Cable cannot succeed where cities "overregulate." Cable cannot deliver the "services of tomorrow." Cable cannot even be considered for some cities.

Ralph Lee Smith, an early popularizer of the concept of the "wired city," expressed disappointment and alarm at the prevailing trend toward concentration of ownership in the cable industry. He warned that "this situation does not serve the interests of city governments. . . . Safe from serious competition, these few companies can and do make unimaginative or limited proposals to the city."[5] Smith was also concerned for the *pattern* of concentration: ownership is increasingly in the hands of other communication media, including producers of programming and equipment manufacturers. This incestuous relationship is a direct threat to diversity of service offerings, and it is sure to grow with further concentration.[6]

Smith's worries are supported by the findings of several Congressional investigatory committees. A 1973 report of the Senate Governmental Operations Committee, *Disclosure of Corporate Ownership*, indicated that the FCC had virtually no accurate knowledge for performing its corporate oversight function: in 1972, 12 banks held controlling interests in no fewer than 20 major broadcasting companies, in violation of FCC regulations holding investment to no more than one percent in any company by a bank.[7] When it was disclosed that many banks held as much as eight percent and nine percent of the voting stock of some broadcasters in their own names (and more in "street names," that is, anonymous accounts), the FCC meekly reacted by raising its prohibition to five percent.

These financial institutions do not play a benign role in the management of cable operators. On the contrary, according to economist Bob Litterman, the major institutional investors–Chase Manhattan Bank, Morgan Guaranty Bank, Pittsburgh National Bank, First National Bank of Chicago, and the Aetna, Allstate, Travelers, and John Hancock insurance companies–"with large blocks of stock, interlocking directors, and huge loans outstanding, . . . play a large role in the direction of the media corporations."[8] For the great majority of cable operators, who are required to seek capital from the large financial institutions, basic criteria established by the lending institutions–besides outright participation in management–severely circumscribe the ability of the independent operator to respond to local service needs.

For example, an officer of the John Hancock Insurance Company, in discussing the avenues available to small operators in locating capital, has advised that there are "three keys to keeping money interested"–be conservative and reduce risks; reduce the risk of the project (before actual operation begins); and either increase the lending interest rate or allow the lender to participate on terms.[9] Not coincidentally, an analyst for Sutro & Company, a large national brokerage firm, felt confident recommending Warner Communications to its clients because Warner's "present philosophy dictates that systems will be developed only on the basis of what cable can now deliver, ensuring that only economically viable franchises will be built."[10] At the time of the recommendation, Warner had just pulled out of franchises in Dayton and Birmingham (leaving two minority cofranchisees in default), had curtailed construction of two-way systems, and had restricted local-origination and public access programming (eliminating the Alternate Media Center's innovative project in Bakersfield, California) to systems "where they appear to make sense."

Is cable, in its present form, really viable? Although U.S. firms are not required to divulge their revenues and operating costs, Canadian firms must do so by law. A 1969 Canadian survey revealed that Canadian cable firms, even in the prepay-cable era, received a return to capital of 12.1 percent.[11] (Martin Seiden, who quotes the report, notes that this figure is below that received by broadcasters. But broadcast revenues and profits have become notoriously inflated in recent years, with television commercial time so scarce and advertisers so many. The point to be made is that 12.1 percent is quite respectable, exceeding the return obtained by AT&T and other blue chip industries.) The inception of higher subscriber rates and pay-cable can only have improved the situation of the cable operators.

Still, to maintain this rate of return requires that expansion not proceed too far: affluent potential subscribers decrease in number as cable descends the surburban enclaves and hilly redoubts, raising the marginal cost of new subscribers well beyond the tolerance of most investors in cable television companies. As Hauser prophesied, "The industry will have to build basic systems, then add on pay-cable and the rest, getting the subscribers to pay for each as it's installed."[12] This "pay as you go" philosophy of conservatism reflects more than sober fiscal policy: in the words of the Network Project: "Cable's domination by (the corporate) segment of society is only encouraged by policies which imply that those who cannot afford cable's costs are not entitled to its benefits"[13]–ignoring the public costs of regulation shared by all and the lost potential benefits of a universal cable utility still unrealized.

Aside from the multitude of technical gadgets made possible by cable–home fire alarms, metering systems for water, shopping by cable, working by cable, and more–popularized in Smith's *The Wired Nation*, there are more substantial and pervasive benefits to the universal cable utility. The Study Team on Citizen Information Needs of the Los Angeles City Citizen Task Force on Cable Communications claimed that the city could "in the long run . . . be highly organized around 'electronic highways,' just as ancient cities were organized around water routes, and more modern cities around land transportation."[14] The Study Team felt that the city could satisfy at least four broad categories of citizen service needs:

1. Public communications: Cable systems in the city of Los Angeles should maximize communication between the city government and the citizen as a facilitator of the democratic process.

2. Communication-transportation substitutes: In recognizing the unique transportation problems in Los Angeles (but similar to problems in most major American cities), careful study should be made of all possibilities of alleviating transportation problems by the substitution of communication.

3. Education: For a number of reasons, long-range planning should be done jointly with long-range planning in the city public school system.

4. Cultural diffusion: The eventual maturity and strength of Los Angeles may be dependent in large part upon commonness of goals of the many individual communities that make up the greater county area.[15]

A complementary report of the same Task Force's Study Team on Franchising Districts recommended the establishment of a citywide cable network, with channels franchised by planning districts so as to encourage the greatest democracy and optimal level of diversity in programming; with the city making the very substantial investment in construction of the physical plant, local communities and citizens (as well as commercial data and production services) could concentrate their limited resources on the provision of programming and services.[16] A most important aspect of this recommendation was the suggestion that the city *not* carry rebroadcast signals, since private cable television operators already supplied such service where required, and since keeping the system a common carrier would largely remove it from FCC jurisdiction according to a recent federal appeals court ruling.[17] Channels could be preserved for services required and requested by the various residential and interest communities, rather than dissipated in duplication of available broadcast programming.

A Delphi survey (a forecasting methodology) of experts in the area of broadband communications, conducted by Paul Baran, revealed a collective opinion that within five years of the introduction of nonbroadcast, two-way cable services, an operator could realize revenues of approximately $20.10 per household per month over and above normal subscriber fees–or about $250.00 per year.[18] Of course, not all costs would have to be carried by the subscribers; tariffs could be charged suppliers of some programming–for example, banks using a channel for electronic banking–and the revenues used to defray subscriber costs. Or the city might decide to subsidize certain services and certain subscribers outright, considering the public commonweal enhanced by accessibility of these services: allowing the unemployed to access a "help wanted" program at reduced or no cost might be preferable to having the unemployed remaining dependent upon unreliable interpersonal grapevines for locating jobs, using public welfare to survive. Ultimately, the matching of needs of all kinds with resources for their satisfaction–made possible by better and more accurate flows of information in the city–might be found to be so desirable as to have the entire system subsidized by the rejuvenated municipality.

CABLE AS A COLLECTIVE GOOD

Cities have traditionally drawn human beings together in productive enterprise. As a result of their constitution and purpose, cities demand participation of their inhabitants in the social interactions of urban life. The provision of required quantities of valid information for their citizens, as well as opportunities for their citizens to initiate the transfer of information to others, is a service that the cities must provide as surely as they provide physical security and physical mobility. If the cities of tomorrow are truly to be built around "electronic highways," then the creation of those highways is as valid a public task as the construction of freeways and transit lines.

In order to provide public services for their citizens, cities are granted powers of taxation and management by the states, which charter them into existence. The power to provide "collective goods" is a special power that allows the cities to enter into the business of cable communications.

Alan Rabinowitz has synthesized this definition of "collective goods":

". . . the functions of local government from which no private entrepreneur could make a profit, for which prices would not reflect real value in use, and for which costs do not rise as one extra member of the public partakes of the service."[19]

Faced with the reluctance of private enterprise to construct two-way cable systems, cities could conclude that cable is truly a collective good. The cost of adding subscribers to a citywide system would be extremely small per subscriber and could easily be offset by raising tariffs for vendors receiving the benefit of additional potential customers (or by simply returning the cost to the subscriber as an installation fee). On every criterion, cable communications rates high as a collective good that the cities can provide: there is no latent private competition; social value would far exceed the costs of operation; and costs for adding subscribers would not necessarily increase the general costs of system capitalization and operation.

It is important to confirm cable as a collective good, for municipalities may only incur debt to underwrite bona fide public purposes.[20] Until very recently, there were few convincing examples of cable television being operated as a public enterprise serving a collective good. Private operators, furthermore, consistently downplayed the success of the small municipal cable systems already in operation; fighting regulation as public utilities, the private operators attempted to define "cable"—the *potential* technology, as well as their own limited one-way entertainment medium—as a luxury better left to the private sector.

But researchers and theorists have recognized the potentialities of a non-entertainment cable medium in the third decade of the cable medium, beginning with the work of the National Academy of Engineering in 1969[21] and Baran's analysis of the desirability of two-way services in 1970.[22] An article by Edwin Parker and Donald Dunn in 1972 suggested that cable could solve many pressing social problems—for instance, inequality of access to education.[23] As a result, a number of studies were initiated in 1972, including a MITRE Corporation two-way hardware experiment. This work was motivated by a concern expressed by MITRE scientist John J. O'Neill:

> I think there are a lot of public interest applications in services that may not be addressed by a private entrepreneur. . . . And the cities are spending fortunes now, literally, doing these types of functions, and they have got to become more productive at it. Telecommunications delivery of these services can help to do that and increase that productivity quite a bit.[24]

A sophisticated feasibility study recently performed on behalf of the city of Portland and Multnomah County, Oregon, proposed two overlapping cable systems, one commercial and one noncommercial.[25] The study claimed that "particular emphasis should be paid to non-broadcast services to the subscriber," since "broadcast services alone may not lead to high subscriber penetration."[26] High penetration would be required, not to maximize profits for the operator of the commercial channels but to ensure that public services, "urgent in some places," received the widest possible distribution.

Michael Molenda has compared the effects of the arbitrary designation of conventional cable system boundaries with the consequences of the arbitrary construction of school districts: inevitably, as children in richer districts receive a higher quality of education than those residing in poorer districts, citizens residing in affluent residential communities will receive greater access to useful information resources than those in less affluent or "flatland" communities as a virtue of their attractiveness for cable entrepreneurs.[27]

If cities are to prevent further uneven social development–compounded by disparities in educational and economic informational resources–they will have to realize the public worth of cable communications and work to convince their legislators and citizens accordingly. Only when cable is seen as a public good, rather than as a luxury, will resources become available for its development. When cable is ruled a public good, it can be operated as a public enterprise, even as a "yardstick" to measure the quality of current service provided by private corporations.[28]

One reason for caution on the part of public enterprise champions, however, is the astonishing inefficiency of many federal public enterprises, most notably the Postal Service and the Department of Defense (though few agencies are untouched by waste.) It may be that there is simply a certain limit in size beyond which all enterprises become unwieldy and wasteful. As they become distended and removed from direct accountability to their clients, their ability to serve may be overcome by overbureaucratization and internal empire building. Mark Green has made a convincing case that limiting the size of all enterprises, private and public, would greatly increase the efficiency of American business and government and improve the general national economy.[29]

Assuming that the public cable system could be limited in size so that it could be efficient, would it still be accountable? Would it engender real democratic participation by workers and consumers in the affairs of the city? Care must be taken that any public enterprise *is* democratic and accountable. As Herbert Gans has advised, ". . . state ownership or control must always be preceded by the adjective 'democratic,' and the kinds and degrees of economic democracy needed must go beyond the limited amount of democracy that America has achieved in the political sphere."[30] In order to ensure the ability of a public enterprise to start out "with a clean slate," Gans proposes the isolation of state ownership from "the present political economy."[31] To function properly–that is, to fulfill the needs of the urban masses–the cable enterprise must not be forced to operate with political and economic constraints exercised by exploitive special interests.

Cable communications, fulfilling all the qualifications for a public good, can be operated as a public enterprise funded by public debt. More than a counter to private enterprise, it can be an efficient means for converting public energies and resources into communication facilities to improve the quality of life in the cities.

THE POSITIVE ECONOMICS OF PUBLIC OWNERSHIP

A city must examine the economic viability of any public service it undertakes. The "bottom line" for the city might be radically different from that of the commercial cable operator; nevertheless, the city's management would have to consider a decision to develop public cable communications very carefully.

However, before the city rushes off to one of many "cable experts" for advice as to how it should proceed, it should heed a warning from attorney Don LeDuc:

> Studies in cable history and cable economics offer an interesting contrast in source problems, for while the medium's two decades of transition are generally covered in a pro forma "out of the foothills of Pennsylvania came community antenna" paragraph, its economic implications are often analyzed far too minutely by researchers seeking legal rather than scholarly evidence. . . . [F]indings, no matter how scrupulously derived, will often be influenced by the funding organization.

And further,

> . . . divergence among studies employing similar analytical techniques appears to indicate that efforts to locate a single determinant within a process [of cable development] this complex and irrational still remain beyond the present state of the art of economic analysis.[32]

Expert advice should be sought with an understanding that, ultimately, the final decision to build or not to build will have to be made by an informed municipal government and citizenry.

(Rabinowitz has criticized an anachronistic call for municipal fiscal caution—possibly due in large part to New York City's very particular situation—attacking these "abstract propositions" and "call[s] for symmetry and moderation" at a time "of vast intermetropolitan and intrametropolitan upheaval" and "expansion in the scope of government projects."[33]

Michael Herrero, discussing the development of community in the "new towns" of the sixties, has observed that the towns themselves do not possess the population necessary to fully utilize an expensive telecommunications infrastructure.[34] Rather than capitulating to this situation, Herrero suggests that the cities involved should receive assistance from the Department of Housing and Urban Development developing an urban growth strategy, which would create new community programs emphasizing systems experimentation in telecommunications and other fields.

Herrero's courage and action-oriented philosophy is hardly compatible with the mode of leadership practiced by traditional public administrators. A danger lies in putting too much faith in the ability of local administrators to use federal support for projects at the local level. John Renfuss notes that "the infusion of federal monies into state and locally administered programs has not brought relief to urban problems–indeed, sometimes it seems to exacerbate them.[35]

Bringing responsibility for financing and managing the operations of a publicly owned cable system back to the city–while making its citizens cognizant of their role in making the system a success–will be a necessary precondition to the fullest integration of telecommunications infrastructure and social progress.

How can the cities assume this responsibility?

First, the cities can draw upon their general funds and operating finances to provide the necessary seed money for system construction. This has been the means by which all existing publicly owned cable systems have been constructed. In the city of San Bruno, for instance, municipal water monies were loaned to the public works department to finance the first phase of development in that city. This method works especially well in small cities, where surpluses exist and system operations can generate sufficient revenues to repay interdepartmental loans.

Unfortunately, few large cities still retain surpluses in the municipal treasury. With declining tax bases, these cities must look to outside financing to build their systems.

A second alternative is for the cities to issue municipal bonds. *Revenue bonds* are pledges made to investors to repay the price of the bonds, with interest, out of the operating revenues of the completed public project. Since few existent cable systems have demonstrated outstanding profitability, investors are unwilling to underwrite potential offerings. One analyst has predicted that were investors to be found, the conditions imposed by those investors might be prohibitive–higher interest rates, possibly, and a shorter maturity date for outstanding securities.[36]

General obligation bonds pledge the city's taxing powers against the repayment of principal and interest due the investor. General obligation bonds are less difficult to market when a city is developing a new service, such as cable communications; however, the actual success of the offering is dependent on the debt situation of the issuing city, as well as the city's past record in retiring debt.

Additionally, general obligation bonds require, in most municipalities, at least a majority, and often a two-thirds, approval by the voters. Since the debt is assessed against municipal taxes, the citizen must bear the burden of securing the offering.

The third alternative available to the cities is the creation of municipally endowed *nonprofit corporations*. These entities also would require support from the general funds of the cities, extending the tax liability of the citizens—and thus requiring their formal approval. The corporations would be empowered to issue their own securities, though these would not be attractive until a system was in full operation and demonstrably capable of repaying accumulated debt out of revenues.

(Some professional analysts argue that public support for cable systems is not sufficiently strong—that "it would be hard to justify an expenditure on a CATV system of dubious profit potential, or at least a long payback period, to the exclusion of more important housing or educational projects.[37] These critics describe an unacceptable "Catch-22" situation for the cities: cable cannot be introduced into the city until it proves itself in the city. The responsibility—and the capability—for initiating needed urban services is vested in the city, however, regardless of the doomsaying of distant "analysts."

What advantages are there for the city financing its own cable system? If it is capable of internal funding, it saves—totally—the costs of issuing and marketing debt, including promotion and interest due on principal. It could complete construction quickly: saturation of the entire city could be achieved without sporadic interruptions occasioned by lackadaisical bond sales. This is obviously an optimal course of action for the city in a favorable economic situation.

Even for the city in a less-well-endowed state, the issuance of bonds could have benefits to recommend this course of action over reliance on private capital. The city would realize, through the issuance of tax-exempt municipal bonds, savings that a private operator would lose in the issuance of standard commercial debt. Since interest on municipal bonds is tax exempt, investors are willing to accept a lower yield (or rate of interest) than they would be were they required to pay interest on capital gains, as they would be with commercial debt. If the city were to offer bonds at 8 percent interest, rather than the 10 or 11 percent private operators must offer on their debt issuances (and pass on to their customers), it could realize significant accumulative savings over the ten or more years the debt would have to be serviced—amounting to many tens, or even hundreds, of millions of dollars. Further, autonomy of the system's management could be preserved (at least theoretically), since bondholders would have fewer prerogatives to interfere in system operations than would stockholders or bondholders in a private corporation. Nevertheless, it would still be important for the city to insulate the management of the system from the control of bondholders.

(Another caution is the debt ceiling placed on some cities by their states. Special legislation might be required to secure greater bonding capabilities in larger cities.)

Several states have passed, or are considering, bills that would enable counties to "subcontract" for their constituent cities (the Lakewood Plan, already discussed). If cable were to be developed by the county or authority, the tax burden could be distributed over a wider population and so equalize system costs among client cities. The county could hasten construction in all its constituent municipalities and communities, effecting savings through rapid and large-scale construction. County bond offerings, however, would have to be extremely large, preventing quick sale of the entire lot—and, of course, removing the power of finance and oversight from citizens at the immediate local level, the raison d'etre of the publicly owned cable system for the city.

The city contemplating the establishment of a nonprofit corporation to organize a publicly held operation could consider issuing ***industrial development bonds***, supported with federal funds.[38] Disadvantages of these bonds include involvement with the federal bureaucracy, a $5-million limit on bonding power, and a provision that monies thus obtained cannot be expended for several years preceding or following sale of the bonds. But the city could place the bonds in escrow, rent the system until the debts were removed from the ledger, and take full depreciation for the entire period. This strategy has not yet been used to finance a publicly owned cable system.

A new technique for generating capital is the issuance of "mini-bonds," recently advocated by California Public Utility Commissioner Leonard Ross, among others.[39] These minibonds are small-denomination securities, which closely resemble U.S. Savings Bonds—and indeed, the markets are identical. Local clients of utilities are invited to purchase a share of the debt issued by the utilities. (Ross reports that a power company in Florida received assurances from thousands of its customers that they would lend the company over $46 million, in amounts ranging from $100 to $200,000.) The benefits are two-fold: first, the customers as investors are indirectly involved in the management and operations of the utility that serves them; and second, the influence of distant financial institutions is significantly reduced, encouraging the utility to become more responsive to local consumer needs. Because the large institutions thus lose their monopoly control over investment capital, they cannot enforce high interest rates for lending; the utilities, and their customers, can save millions of dollars with each offering of debt.

An experiment in the formation of a cooperative to provide cable service in rural Wisconsin demonstrates how the minibond concept might be applied to a publicly owned project, albeit a nonmunicipal one.[40] Since private operators could not economically service the residents of Trempeauleau County, citizens there formed the Western Wisconsin Communications Project. Each of the potential 9,500 subscribers will be offered the opportunity to purchase a share of the system. Initial capital for this noncity venture is being sought from the Federal Rural Economic Assistance Program—to underwrite early organization and construction of the system.

(This cooperative, unlike those proposed for urban communities, can count on assistance from local electrical and telephone cooperatives willing to share an existing physical plant. A city would have to develop its own plant. However, most cities already possess their own basic plant—sewers, conduits, transmission sites, and other components—which could easily be adapted to provide carriage of cable throughout urban areas.)

In most communities, the city will be the *only* entity capable of effectively operating a well-financed cable system; moreover, as researchers at the Yale Law School have observed, municipal management could be "construed as involving state action," thus permitting "any unrepresented minority group to protect its constitutional interests in court."[41] With a private operator, there would be no such recourse; until the Federal Communications Commission rules on a citizen complaint, the matter remains in abeyance. Rejection of a citizen complaint by the FCC—which has been the rule in broadcast matters—can only be challenged in the Federal District Court for the District of Columbia, necessitating expensive preparation and further delay and rendering any resolution moot in terms of the original grievance.

Development and construction of a citywide cable communications system will require large transfusions of capital, regardless of who initiates such development. Few operators, even without the spur of city competition, have been able to overcome the economic inertia retarding expansion of their urban systems. Only the city—attempting to provide a necessary public good—has the motivation to secure large-scale financing; as an agent of its citizens, it can realize significant savings in terms of taxes avoided, interest rates held down, and rapid construction productive of prompt cash inflow and reduced debt.

In addition to the economic benefits of municipal ownership must be considered the opportunities it would provide for citizen involvement in the financing and support of the system. With citizen financing comes citizen interest in the care of the cable investment: Is it fulfilling its purpose? Is it being used efficiently? The concept of the minibond is especially well suited to this goal, since it would reduce distant private control over local policy by eliminating the power of monopoly capital in the loan market.

IDEOLOGICAL RESISTANCE MEETS SOCIAL REALITY

Despite all the arguments favoring municipal ownership—or, perhaps, because of them—the ideological critics of municipal investment in cable communications have become particularly vociferous in recent years. Not unexpectedly, these proponents of "free" enterprise found allies in a Republican administration favoring the reduction of public restraint on private capital. The agencies of the institutionalized assault on the potential public role in the development of cable communications were the Federal Communications Commission and the Office of Telecommunications Policy.

The FCC was originally created as an arm of the Congress, but it has since become a tool for enacting administration policy. All appointees to the FCC are nominated by the President, and subjected to only cursory examination by Congress prior to confirmation. The immediate past chairman of the FCC, Richard Wiley, presided over an FCC repeatedly accused of collusion with various communications industry representatives.

In a letter from Wiley protegee David Kinley, then director of the FCC's Cable Television Bureau, Kinley quoted an earlier statement demonstrating Wiley's ideological bias as chairman (despite Kinley's disclaimer that "the Commission has taken no firm position on the subject"):

> . . . I am still firmly opposed to municipal ownership of cable–just as I was a year ago. In my opinion, government simply does not belong in an ownership position of a communications business. In a democracy like ours, the free flow of information and ideas would not be well served by placing (the cable industry's) voice under bureaucratic control.[42]

Wiley's deliberate confusion of political pluralism and capitalist economics is revealing: it assumes, on one hand, that private capitalism is synonymous with democracy; and that, on the other hand, government ownership precludes the exchange of ideas. In fact, it is questionable whether private ownership of the media of communications produces anything like a "*free* flow" and problematic whether local governmental ownership is necessarily restrictive.

When the FCC invited comments on its plans to eliminate a 1977 deadline for upgrading of "grandfathered" cable television systems (those in operation before its 1972 rules were propounded)–going so far as to create a "1977 task force" to expedite deregulation–cable operators flooded the agency with plaints of poverty. Without due consideration, and using figures supplied by the industry association, the National Cable Television Association, the FCC abolished requirements for two-way capability and additional channel capacity that would have made possible the provision of nonentertainment cable services.

Alternative economic analyses submitted by the Cable Television Information Center and numerous state agencies, including the city of San Diego, specifically challenged the FCC's approach to regulation, suggesting the need for an independent assessment of the economic state of the cable industry.[43]

(The indifference of the FCC to the desires of the cities has prompted some cities to consider challenging the FCC in a legal action still under informal consideration; the magnitude of the uprising has grown substantially since the Anaheim-Arlington controversies.)

While Wiley and the other members of the FCC remained adamant advocates of "free enterprise" in opposition to municipal initiatives (with

Commissioner Benjamin Hooks a possible dissident), the opinions of the Office of Telecommunications Policy were perhaps more extreme.*

OTP's former director, Clay T. Whitehead, chief advisor to the President's Cabinet Committee on Cable Television, helped draft a document that demonstrated a coherent plan for the decontrol of cable television and the throwing open of the field to the "rule of the marketplace." The report employed the cliched and ill-founded arguments of the cable industry in concluding that

> it would be unwise for municipalities to function as cable operators. For the foreseeable future, cable system ownership will be a capital-intensive enterprise that may well be subject to rapid technological change and associated financial risks. As long as private entrepreneurs are willing to do so, it is almost certainly an unsound allocation of tax dollars for municipalities to underwrite such ventures. Moreover, with a financial interest in the entity being regulated, it would be inappropriate for a local authority to function simultaneously as the "regulator" and operator of the cable system.[44]

The ideological irrationalities of the private entrepreneurs and their governmental spokesmen hardly allow for the desperate plight of the cities. It is as though these specialists in communications matters, like the specialists in land-use planning or the specialists in public administration, have a one-dimensional view of the contemporary American society and its physical and nonphysical infrastructures. They cannot fathom the connections of the one to the other, or at least they will not do so publicly; to admit that the crisis in the cities has roots in the distorted patterns of communication in the city—and then to admit that that distortion is the result of severe economic injustice in the society—would be politically calamitous.

Rather than deal with such fundamental relationships, the defenders of private enterprise prefer to tinker with the technical aspects of regulation, attempting to draw public attention away from substantial questions reflecting on the hegemonic domination of communications in America. Often they succeed, coaxing their attackers into endless theoretical debates and legalistic nitpicking.

These tactics fail completely when confronted by the cities' objective condition, not only as unsuccessful political entities but as settlements where the

---

*Commendably, Wiley did not allow his personal political views to color his official judgment—or perhaps he detected winds of change beginning to blow: in *Consolidated TV Cable Services, Inc.*, FCC 74-742, 47 FCC 2d 1041, 1043 (1974), and again in *Valparaiso Cable Company*, FCC 76-602, 61 FCC 2d 328 (1976), Wiley joined in majority decisions of the Commission rejecting petitions to deny certificates of compliance to municipal cable television systems in Frankford, Kentucky, and Valparaiso, Florida, respectively.

quality of life appears to be quickly declining. What is needed is a new body of theory based upon urban realities and not pseudolegalistic speculation.

Norton Long has taken the necessary long look at the cities' problems; his main thesis begins a lively and altogether novel treatment of today's urban crisis:

> The rhetoric of urban fiscal woe and its revenue sharing remedy would lead one to suppose that there are a multitude of "sick cities" in a healthy nation from whose wealth a healing balm of money can be extracted. But this is scarcely possible in any fundamental sense. The national economy cannot flourish while the local economies that make it up wither. Only because we fail to see the cities as local economies composing the cellular structure of the national economy is this obscure. Can the cities be fiscally sick and economically sound? Such a condition is hardly credible. . . .
>
> The health of the city is the employment and the upgrading of the employment of its people. . . . The contemporary city of the nation state has looked to the nation and market capitalism to insure its economic viability. In an era of rapidly expanding local public goods, these mechanisms have failed. . . . The cure for the disease, if cure there be, lies in the rebuilding of the city as a political community whose members, in sufficient degree, accept a common fate and empower a leadership to shape that fate as it lies in the power of the city.[45]

Carefully avoiding "patent solutions such as mass transit," Long laments the suburbanization of the American city, which leaves the central core a burnt-out cinder, its inhabitants seen as the "dregs" of society. He calls for a new urban economics, one founded on local processes rather than national models, which can illuminate the cities' plight and suggest plausible solutions. Long refuses to admit that the city is dead, though it displays the symptoms of disease; rather, it has been allowed to "inevitably decay," in line with the pronouncements of suburbanite scholars who long ago fled.

Long has his own prescription for the city, and that is to utilize the energy of the great numbers of people who still choose or are forced to live in the city, to have them exert control over entrenched leadership and inefficient unions of civil servants and unite the various groups of people living in the city in a common effort—to have the city serve its constituency. But the situation is desperate:

> One might consider the outlook for the city as grim and well-nigh hopeless. What could lead one to believe that a business such as a city with so feckless a management, such small incentive to make painful, difficult changes, such powerful public-employee unions, and apathetic public, divided blacks, a barely involved business

> community, a middle class that has flight as an option, and a media whose role is critic can be restored to solvency?[46]

Long demands that we ask questions about traditional (and empirically unfounded) explanations for urban atrophy. In proposing his solution for urban rejuvenation, Long asks another question:

> Of all those with power and a rational self-interest in the city's future none have more than the public employees whose own future and that of the city are hopelessly interlocked. Can they be made to see their interest in the city as their cooperative, a producer as well as a consumer cooperative that offers them in every rank a vocation with meaning, challenge and reward? Can the incentive system of the city unite management, employees, and citizen consumer producers in a humane cooperative to improve the lives of the whole city?[47]

Here Long is at a loss, for he cannot bridge the gap between reality and the ideal; he vainly awaits action on the part of "the labor movement, the young doctors, the public-interest lawyers, the artists, middle class, and the black leaders, who from widely divergent perspectives see a need for a future in the city." Long's excellent analysis erodes into an idealistic muddle, without a means for uniting these "divergent perspectives" or even making them aware of each other.

An urban cooperative, to be successful, would require an urban communications network far more efficient than any now available. It would have to be capable of reaching individual citizens, as well as addressing audiences on common issues. Moreover, it would have to allow all its constituents to interact and to access information from each other and from official repositories of information.

Cable communications could provide such a network. It could infuse the various populations living in the city with a purpose gained through a sharing of perceptions and suggest action strategies for carrying it out.

Long is careful, over and over again, to caution against public services that do not carry their weight; in fact, one of his remedies for the cities is the paring down of inefficient services. But a cable system making possible the variety of communications necessary to the city-saving campaigns ahead is the basic requirement for concerted citizen action—it is the service which makes others that much more efficient.

A sense of optimism, as well as a sense of economic propriety, would be essential to the successful implementation of comprehensive, community-serving cable communications—just as it will be to the salvaging of the cities. Skepticism can be afforded by cynical suburbanites, to whom the city is only a large factory closed with the five o'clock whistle. For those residing in the heart of the urban maelstrom, or even on its fringes, conventional wisdom no longer wears very well. This is a time for experimentation in technological *and* social innovation.

Cable communications, owned and operated by the people of the city, could help create an environment to germinate the most important qualities of a dynamic urban community—self-reliance, cooperation, an *awareness* of the objective urban reality, and a recognition of the power residing within the people to remake that reality in *their* image.

NOTES

1. "In Cable Franchising, the Cards Are Being Played Closer to the Vest," *Broadcasting* (February 4, 1974): 49.

2. "Physician, Heal Thyself," *Forbes* (December 1, 1973): 65.

3. Federal Communications Commission, Docket no. 19988 FCC 74-1279 (released: December 9, 1974).

4. "Cable-TV Dangles New Lures," *Business Week* (December 1, 1974): 64.

5. Ralph Lee Smith, "Ownership Policy and the Cable Industry," *Yale Review of Law and Social Policy* 2, no. 3 (Spring 1972): 271.

6. Ibid., p. 263.

7. "Senate Study Sees Big Bank Presence in Broadcast Firms," *Broadcasting* (January 14, 1974): 41–43.

8. Bob Litterman, "Who Owns the Media?," *Pacific Research and World Empire Telegram* 4, no. 1 (January-February 1973): 14.

9. MITRE Corporation, *Symposium on Urban Cable Television*, vol. 2 (McLean, Va.: MITRE Corporation, March 1973), pp. 21–22.

10. Arthur E. Rockwell, *Quarterly Survey of the Cable Television Industry* (San Francisco: Sutro & Company, January 30, 1974), p. 13.

11. Martin H. Seiden, *Cable Television U.S.A.* (New York: Praeger, 1972), p. 48.

12. "Cable-TV Dangles New Lures," p. 64.

13. The Network Project, *Cable Television: End of a Dream* (New York: The Network Project, Columbia University, 1974), notebook no. 8, p. 26.

14. Los Angeles, Study Team on Franchise Districts, *Interim Report on Franchise Districts* (Los Angeles: Los Angeles City Department of Public Utilities and Transportation, January 15, 1976), pp. 2–7.

15. Ibid., pp. 2–7.

16. Los Angeles, Study Team on Franchise Districts, *Interim Report on Franchise Districts* (Los Angeles: Los Angeles City Department of Public Utilities and Transportation, February 17, 1976).

17. "Court Strikes Down FCC Pre-emption of Regulation of Leased, Two-way Cable Channels," *Broadcasting* (February 16, 1976): 54.

18. Paul Baran, *Potential Demand for Two-Way Information Services to the Home, 1970–1990* (Menlo Park, Calif.: Institute for the Future, December 1971), pp. 6–26.

19. Alan Rabinowitz, *Municipal Bond Finance and Administration* (New York: Wiley-Interscience, 1969), p. 44.

20. Gordon L. Calvert, ed. *Fundamentals of Municipal Bonds* (New York: Securities Industry Association, 1973), p. 19.

21. National Academy of Engineering, Committee on Telecommunications, *Telecommunications for Enhanced Metropolitan Function and Form* (Washington, D.C.: National Academy of Engineering, 1969).

22. Baran, *Potential Demand for Two-Way Information Services.*

23. Edwin B. Parker and Donald A. Dunn, "Information Technology: Its Social Potential," *Science* 176 (June 30, 1972): 1392-99.

24. MITRE Corporation, *Symposium*, p. 155.

25. Telecommunications Management Corporation, *Cable Communications for the City of Portland and Multnomah County* (Los Angeles: Telecommunications Management Corporation, 1973), p. 25.

26. Ibid.

27. Michael H. Molenda, "CATV and Access to Knowledge," *Yale Review of Law and Social Action* 2, no. 3(spring 1972): 243–51.

28. Peter Barnes and Derek Shearer, "Beyond Antitrust," *The New Republic* 171, nos. 1-2 (July 6 and 13, 1974): 16–19.

29. Mark Green, Beverly C. Moore, Jr., and Bruce Wasserstein, *The Closed Enterprise System: The Report on Anti-Trust Enforcement* (New York: Grossman, 1972).

30. Herbert J. Gans, Letter in "The Planning Debate," *Social Policy* 6, no. 7 (May-June 1974): 25–27.

31. Ibid.

32. Don R. LeDuc, *Cable Television and the FCC* (Philadelphia: Temple University Press, 1973), pp. 274–76.

33. Rabinowitz, *Municipal Bond Finance*, p. 47.

34. Michael C. Herrero, *New Communities and Telecommunications* (Chapel Hill: Center for Urban and Regional Studies, University of North Carolina, 1973).

35. John Renfuss, *Public Administration as Political Process* (New York: Charles Scribner's Sons, 1973), p. 78.

36. Robert G. Bulot, of Stone & Youngberg Municipal Financing Consultants, Los Angeles, private communication, April 18, 1975.

37. "County-City Cable Plan," *Broadcasting* (February 25, 1974): 54.

38. MITRE Corporation, *Symposium*, p. 28.

39. Leonard Ross, "Minibonds for Utilities: An Idea Whose Time Has Come?," Los Angeles *Times*, June 2, 1975, p. 7.

40. "CATV of, by and for the People," *Broadcasting* (March 11, 1974): 79.

41. "Toward Community Ownership of Cable Television," *Yale Law Journal* 83, no. 8 (July 1974): 1723–24.

42. David Kinley, of Cable Television Bureau, Federal Communications Commission, private communication, March 25, 1974.

43. "Poverty Is Major Issue in Comments on '77 Deadline," *Broadcasting* (March 24, 1975): 39.

44. President's Cabinet Committee on Cable Communications. *Cable Report to the President, January 14, 1974*; reproduced in *Television Digest* (January 21, 1974), special insert.

45. Norton E. Long, *The Unwalled City* (New York: Basic Books, 1972), pp. xi–xii.

46. Ibid., p. 188.

47. Ibid., p. 189.

CHAPTER

8

WHERE THEORY ENDS

A CRISIS OF RESPONSIBILITY

Before the crisis in New York City catalyzed public opinion and stung city managements into an awareness of the precarious state of their domains, most municipal officials were extremely cautious in their activities not to upset a perceived status quo. Federal and state money could always be obtained to underwrite urban programs designed to forestall cataclysmic failures of the urban system, leading to social unrest and even open revolt.

But, eventually, the states began to fall away. Their conservative legislatures were never sympathetic to the urban masses, with their labor movements and minority rights groups and gay liberationists. Skyrocketing inflation and the general unwillingness of exurbanites to support the cities combined to effectively curtail state aid to the cities. Murray Stedman, criticizing the state-cities relationship, sees only one course open to the cities:

> As a result of this situation, the cities find themselves in a dilemma. On the one hand, they can wait for the solution to their problems until the states enjoy a newfound prosperity, presumably with federal funds under a revenue-sharing plan. Or, on the other hand, refusing to sit tight until the Day of Awakening finally arrives, the cities can take their case directly to the federal government.[1]

Today, as the New York City crisis demonstrates, even the federal government cannot be relied upon to "bail out" the cities: with large portions of the national budget allocated to the production of implements of war and welfare, little is left to invest in improving urban life.

The cities have been thrown back on themselves—and it will be up to present urban managements to work with their citizens in building citywide coalitions able to underwrite the development of necessary urban cable communications. Not only must these coalitions strive to discover more efficient technologies and methods for the provision of urban cable service but they must also enable the citizenry to critique the services currently being delivered and to specify what might be required in the future.

Many cities may find themselves unable to fulfill this responsibility—small cities, because they do not have the resources to conduct such a program, and larger cities, because every available resource is already committed to one or another holding action. For this reason, cities may look to regional organizations to take over any and all municipal planning that might require work in some nontraditional area of concern. Mel Scott, however, sees a much more limited role for the regional authority:

> The metropolitan agency can envisage the overall systems of facilities and services and can indicate, in the main, how areas for urban expansion should be developed, but the more detailed planning which is needed to assure individuals and groups of their rightful place in society is essentially the responsibility of local agencies, especially if the principle of genuine citizen participation (rather than token participation) is to be invoked, as in the model cities program. There will be times, nevertheless, when the metropolitan agency should intercede on behalf of certain groups to protect them from parochial prejudices and infringements and to maintain desirable standards throughout the whole urban area.[2]

Contemporary history has not proven the metropolitan agency to be overly innovative; usually created for one specific purpose—planning of housing, waste removal, water supply, and so forth—it is incapable of activities outside its limited role. It is conceivable that a special "cable district" could bring expertise to bear on the problem of incorporating public purposes in a publicly owned cable communications system. Generally, however, broad authority does not go hand-in-hand with public involvement.

With each reduction in scale, the distance of the citizen from those making decisions—and the leverage enjoyed by powerful interests adept at separating citizens from decision making—are decreased. But, as has already been pointed out, there is a lower limit; isolated communities are financially incapable of constructing a publicly owned system.

Why have cities been so slow to recognize their need for better urban communications, for channels of communication and content based on the requirements of their citizens? Why have they not allied to pool resources and discover common solutions?

The city is, at least politically, a microcosm of the national governmental scene. Officials are elected to office on the basis of support derived from a few powerful groups: voting is merely a public decision as to which power bloc to support. In fact, urban issues seem to generate even less popular enthusiasm than do national issues.

Too often, the urban authority–the council members, the mayor's office, possibly some other city elective officers–identifies itself with small, powerful interests, whose concern is personal enrichment and not the success of the city as a place to live. This authority works to defeat citizen initiatives to improve one or another aspect of the quality of life in the city, and sometimes–through traditional means, such as preferential zoning practices, or through less traditional means, such as encouraging sham "public participation" in decision making–works to degrade the city as a home.

A few cities seem to encompass enough diversity of viewpoints, however, to allow for operational innovation, even if social innovation–like true urban democracy–is still out of the question. The Urban Institute, after studying several cities where technological advances were successfully implemented, declared that the following "elements of success" were present in every successful program of innovation:

> The systems approach was crucial in analyzing [a] . . . department for potential areas of applying technology and producing savings.
>
> A knowledgeable person was in a position to see how technology could be mated with a particular problem.
>
> The cooperation and commitment of industry, even though it was withheld until pressed and persuaded, was essential to the development.
>
> A close working relationship between outside experts and department heads and employees was important for two reasons–first, to gain acceptance of the notion of innovation, and second, to help assure success through the contribution of their operational expertise.
>
> There were innovative chiefs and managers in the department who, under appropriate conditions, were willing to engage in a novel development program.
>
> A demonstrable benefit was presented to the citizens–in this case, improved efficiency of a vital service–and the story was carefully presented and not oversold.[3]

In contemplating a citywide cable system, it may *not* be possible to adopt a pure systems methodology in evaluating basic needs: very often, the systems analyst, in order to conceptualize a problem, defines himself as "objective" and "beyond" the system boundaries. These "outside experts" should play a limited role in the municipal-cable planning process: this is where citizens, both experts and novices, should work directly with department heads and other municipal employees. Together, they *can* generate cable system designs through cooperative planning–with citizens' needs preeminent.

The crucial question, and one not likely to be answered without experience in the actual operation of a citizen-controlled cable communications system, is exactly who should initiate the process of innovation and citizen involvement. So far only municipal officials, with or without explicit community support, have undertaken the creation of municipally owned cable television systems. They have been less than successful in involving citizens in their management activities, though they have been extremely successful in providing conventional cable television services to their constituencies at remarkably low rates.

Two prominent clients, among others, have utilized outside experts in planning for cable communications. Jacksonville, Florida, received preliminary assistance from the Cable Television Information Center, though that activity has not been continued (and municipal ownership remains problematic). The system most likely to soon provide advanced telecommunications capabilities for its citizens (including interactive communications and community-produced programming and services) is that of the Southern California-based Public Cable Television Authority, a joint-powers agency representing the cities of Westminster, Huntington Beach, Newport Beach, and Fountain Valley. With aid from planners John Bateman and Associates, the Authority has been able to design a system attractive to municipal officials and potential investors alike–placing the Authority in the enviable position of being able to choose from among willing underwriters. (Details on these systems are contained in the Appendix.)

Is it possible that in the modern American city, forces lie dormant that could energize similar processes of their own accord? Have the planners and managers misread the citizenry? If there are energies waiting to be tapped, a catalyst will be required to exploit these resources. Whether it comes from an enlightened management or a desperate citizenry will depend on the severity of urban environmental degradation and the level of awareness of those elected or appointed to serve the people of the city.

HUMAN NEEDS, UTOPIAS, AND THE GREAT UNKNOWN . . .

The case can be made that all that has preceded is conjecture, that urban problems are too complex and deep rooted to succumb to improved urban communications–and that in any case, the costs of such a solution prohibit its ever being implemented. It is futile to argue against such pronouncements; in fact, in some ways, there have been too many grandiose plans already, souring the public on all schemes that promise to rejuvenate the urban environment.

There is, for instance, the sad history of MXC, the "Minnesota experimental city." MXC was to be a planned city of 250,000, incorporating all the best contemporary ideas in urban design. Kas Kalba, relating some details of the plans for MXC, confessed that the cost of the city's "built-in" cable system would be $300 million, nearly six percent of the "typical infrastructure expenditures" for new communities; this extreme cost (the telephone system would cost only $125 million, according to Kalba) would be "justified by the integral part that telecommunications will play in the community."[4] Not only would it "make possible land-use innovations as well as innovations in the fields of education and health care delivery," but it would also enhance life-style options and overall quality of life"; and, finally,

> it should be remembered that the economic base of the city will be telecommunications-oriented manufacturing and service industries. Although no single one of these factors may justify the implementation of such an elaborate telecommunications infrastructure, taken together they do.[5]

Having laid out the blueprints for this major component of the new MXC, Kalba warned that

> current surveys . . . indicate that the population at large is not interested in most of the advanced communications services that telecommunications engineers and planners often rave about. It is consequently critical that interaction between MXC community representatives, real or surrogate, and ideas about future communication services . . . start as soon as possible.[6]

Such calculated prescriptions for urban success cannot be readily transferred from the unbuilt model to the real city. There is no telecommunications infrastructure in Los Angeles, or New York, or Cincinnati. There are only angry citizens who will shudder at the thought of $300 million in new taxes, unless they are involved in planning from the moment system design begins.

The situation is not as bleak, however, as Kalba's "current surveys" would indicate. Martin Kuenzlen believes, contrary to the survey-oriented planners, that

> needs cannot be discovered or rendered explicit . . . with isolated, randomly selected interrogations, surveys and statistics, nor with computer sensors and citizen feedback systems which accept and reproduce every kind of irrationality. They lead either to distorted or banal results; results which simply reflect the needs which are

> manipulated by the capitalistic system. Potential human needs, including the need to experiment, remain repressed and do not appear for the majority of the people.[7]

Harold Sackman suggests engaging a wide variety of community members in a planning effort directed by the theory of mutual expectation, so that

> the planning process is initially triggered and created by an effective consensus [what Sackman terms a *planning community* or *planning public*] in the concerned community, a consensus which is shaped by mutual expectations concerning social values, group goals, resources, and priorities.[8]

In this way, the full range of possibilities for a "mass informational utility" can be accurately determined, taking into account every relevant variable and supplementing any one individual's limited knowledge about his or her community's needs.

Erasmus Kloman, reflecting on the recent surge of interest in "technology assessment" as a planning technique, believes that "the emergence of technology assessment as a tool or process of decision-making reflects an understanding that the knowledge and wisdom required to inform decision makers does not reside in a single place."[9] Yet, as Daniel Roman critically observes,

> The significant point is that decisions to push technology can and often are made unilaterally where the prime motivation is provided by either the government, industry, or the technologists; or, the decision can be bilateral with two of the aforementioned three interest groups in partnership; or, all three groups may be parties to a trilateral agreement. In recent years it has become increasingly apparent that the technologists, government, and industry do not have the sole right nor sole domain in determining the direction of technology.[10]

In the face of this continuing contradiction, Roman has this recommendation: "The government as coordinator is no guarantee of objectivity, but the government acting as a focal point should provide support and decisions on technology promotion consonant with the greatest good for the greatest number."[11]

There is no substitute for vigorous and informed public participation in, and direction of, plans to implement a publicly owned cable communications system—nor is there a convenient terminus for this activity. The ability of the urban masses to learn to govern themselves, to provide capital and labor to satisfy their own needs, and to shepherd these investments is at least as important as the success of the cable system itself. Abandonment of the operating enterprise to a professional managerial elite would signal a very real failure on the part of the participants and the city in general. As Robert Dahl has so astutely noted:

> The old liberal conception of a privately owned enterprise consisting of freely contracting parties operating essentially without hierarchy and power relationships was virtually all myth. . . . Yet socialists who hoped hierarchy would vanish if ownership were transferred to the state, or to a disembodied entity called the people, were profoundly disillusioned to discover that, far from ushering in industrial democracy, government ownership merely preserved or even strengthened hierarchical authority.[12]

Even the elimination of hierarchical authoritarianism is not sufficient, alone, to promote social justice within an urban population joined by an electronic "nervous system": with the defeat of centralized (and unworkable) hierarchy, the city's masses must be wary of new exploiters arising within the changed environment. Kuenzlen has warned against "power elites expressing . . . their functional ruling power and dominating, with it, the environment."[13] It is impossible to foresee the challenges to new freedoms (and responsibilities) that will occur in the cable-connected city; it is possible to be ready for their appearance.

In one sense, it will be difficult to "plan for" the future: every action of the urban masses will actually be creating that future as it occurs. This, however, is a cause for optimism for, as Alexander Tzonis assures, "the pessimism of ever achieving a nonoppressive environment reflects the fact that the authoritarian society has exhausted its ability to invent new products, to identify new objectives and new utilities."[14] Tzonis' plan is for movement, for commitment, and for the overcoming of obstacles in the way of a nonopppressive environment–obstacles attributable to the false construction (both physical and ideological that make up the contemporary urban environment:

> Only direct experience of relationships between men can give the taste of a liberated environment. This is where theory also ends, confined to describing only the abstract organization of such a society. For this reason, as rational design had to introduce verification, incorporating the scientific method, to achieve its objectives, postrational design has to introduce action that affects social relations in its methodology to accomplish its new objectives. Because the bitter or sweet taste of relating to man in oppression or liberation can be found only in action.[15]

Unlike the typically one-way corporate media of today, cable communications could enable men and women of the city to have something very much like "direct experience" with others "in action."

Yet, there remains the unanswered question of who will initiate planning for a cable communications system. There is evidence that, in some instances, local citizen movements can encourage divisions within hardened municipal bureaucracies, into which agents of innovation can enter; then, through the simultaneous diffusion of nontraditional ideas and the attrition of "old hands," the bureaucracy can begin to reflect the dynamic quality of the active citizenry.[16] How often this is actually the case, and how long new attitudes and modes of behavior can prevail, is problematic. Few cities appear to have reached this climax stage of development. Instead, their managers (and the interests they serve) continue to rely upon the ignorance of the people to sustain an essentially oppressive and inefficient method of government.

Yet, the cities *are* in a state of social and monetary crisis. In this volatile condition, the introduction of advanced technological innovations could have dramatic consequences. As Franklin Huddle has observed,

> society is confronted with the most radical force for change in all history. For its own protection, it would seem that society should exercise some extent of purposeful control over the effects of this radical force, permitting, encouraging, even seeking out, the good, while reviewing, retarding, even inhibiting the hazardous or faulty.[17]

Strategies must be devised to ensure that cable communications, with the potential for altering the social relations within the cities, can be brought to the cities in a way that encourages the realization of progressive social ends, which aids in the restoration of the cities' fiscal health—and which promotes the establishment of democratic control of the cable communications system itself. Such strategies may seem, without precedent, nearly utopian—but, in fact, they are being formulated in American cities today.

STRATEGIC ORGANIZING

The ideology of contemporary Western society, stressing individualistic self-reliance and scorning collective behavior, is an obstacle to the implementation of a "democratic" cable communications system. The conflicting demands made upon urban residents—often coercive in nature and a result of dependencies on distant corporate entities (employers, vendors, and so forth)—create "communities of limited liability," in which each resident only makes those commitments to the local community that are necessary to maintain personal security in the face of threatening external forces.[18]

Organizing under these conditions is challenging—and often productive of no change or of unexpected change. The level of public commitment to the development of municipal cable communications systems—indeed, the public's awareness of its own collective needs—is dismally low. The time when cable-carried polling services can be "used" primarily to provide feedback to the information utility designers and developers to aid them in providing services desired by users still seems remote.[19] Perhaps this is why the advocacy of cable communications development has been restricted to a "vanguard" of video enthusiasts—artists, producers of local programming, and the like—and technologists.

The advocates are in a precarious position. On the one hand, they must wholeheartedly press for acknowledgment of the worth of a cable communications utility in the hope that even the slightest recognition of its ultimate potentialities will merit its consideration for a place on the public agenda. On the other hand, they must guard against overoptimism and the inevitable disappointment that follows naive and unsuccessful political initiatives. Seymour Sarason captures the poignancy of the advocates' dilemma:

> Those who conceive the program, passionate and committed to a set of substantive ideas they consider innovative and needed, are not in the best psychological set to deal ahead of time with the radiating community consequences of their program. They are psychologically in a set in which optimism and hope and possession of their version of the truth obliterate a systematic approach to the question.[20]

Even if the visionaries do possess a "sensitive awareness to these questions," one

> can predict that they will encounter in themselves two sources of resistance. The first is that it will take time, more time than they are willing or prepared to give. The second source is that as they tap into existing networks to get soundings, they are likely to find—they will always find—nodes of opposition. They are, in fact, likely to find that their ideas can be expected to galvanize existing networks, near and far, in opposition to the program.[21]

An obvious source of such opposition is the commercial cable industry, which knows its history of service is sure to suffer in comparison with the limited but impressive achievements of public cable operations. Another source of resistance, part of the same "network," is found in the host of cable regulators (at the local, as well as the federal, level), whose goal is to maintain the status quo, which favors the "orderly" exploitation of the new medium by the commercial industry.

The best-informed opposition will come from within the ranks of the advocates themselves, enunciated by researchers and planners wary of the public's presumed inability to effectively use a two-way cable communications system. One researcher has gone so far as to recommend a moratorium on the development of a "community information utility" until the year 2000: "During the interim, we may expect both supplementary and complementary technology to mature, and we may expect the general level of understanding of computers to rise, particularly if we actively work toward that goal. And perhaps our society will be more sure of itself by then."[22]

But our society cannot afford the luxury of waiting for communication-related problems to work themselves out. As Hans Enzensberger warns, "collective wishes"–such as "the need for new forms of interaction" and "the need for self-determination"–"capital often recognizes more quickly and evaluates more correctly than its opponents but naturally only so as to trap them and rob them of their explosive force."[23] So cable communication is reduced to "access" and the potential of cable to the carriage of pay-movies. The year 2000 will be disappointing in its sameness unless the advocates of public ownership take the risks necessary to build a broad base of support for the design and implentation of public cable communication systems.

Building this broad base of support will not be easy. The attachment to place and persons that characterized early communities is missing in the metropolitan environment; as Jacqueline Scherer has remarked: ". . . I see that my communities are not as permanent as those of my parents. I am not at all certain that my community of today will be the same as that of next year, and I am almost certain that the community of my children will be different from mine in the future."[24] This personal discontinuity and alienation signals the end of commitment to the collectivity–and whatever the current shrunken state of community might be, it is the vital link between the individual citizen and the larger administrative entities that must be compelled to fund and implement the development of the public cable system. Exploring the possibilities for innovative forms of government inherent in "new towns," Royce Hanson identifies the community–the local collectivity–as the basic building block for urban government:

> In a new community, you can tier your governmental organization in such a way that each neighborhood within the town has a system of government based on delegated powers, and at the same time have a townwide central government. In fact, the tiering of general government is one thing we should definitely experiment with, in new towns and elsewhere. . . . [T]here is no reason why there cannot be, in each community or village, an organic structure of government which is fairly general in its scope of activities, but which permits people at the village level to deal with the village aspects of government.[25]

These local "organic structures" could subsequently join to form administrative entities able to cope with intercommunity problems without sacrificing local responsiveness and accountability:

> The system should permit various minorities and majorities to be represented, and to have their interests aggregated differently at

various levels. They could combine in various ways to deal at a larger level with problems that simply cannot be handled at a smaller level, and vice versa.[26]

New towns cannot be created out of old cities—but new *communities*, along the lines suggested by Hanson, could be identified and strengthened by the operation of a public cable system, which would make them responsible for the management of local "loops" and require active citizen participation in a central oversight directorate. In fact, such a plan was recommended to the city of Los Angeles by a citizens task force, empowered by the Board of Public Utilities and Transportation of the city to help generate regulatory policy.[27] High-capacity trunklines (possibly incorporating waveguide or optical-fiber technology) would connect nearly two-dozen local subsystems. The trunk lines would carry (1) citywide signals, such as educational programming and municipal-information interactive communications, and (2) intercommunity signals, as agreed upon by the participating communities.

The local subsystems (proposed to be coterminous with carefully designated planning districts) would be under the autonomous direction of local electoral or participatory bodies; except for citywide signals, communities could request or exclude any vendor of services—including special programming, commercial services, or communications from other subsystems. Citywide signals would operate under the jurisdiction of a central authority elected from among the various subsystem communities.

Local communities could thus prevent an "invasion" by external entities—for example, national banks offering electronic fund transfer—which might threaten indigenous commerce or public enterprise. On the other hand, the same communities might work to secure the services of a vendor of electronic fund transfer technology in order to help local financial institutions renovate their own services—or, in line with this example, they might contract with a financial consulting firm to set up a community-owned credit union operated "over" the cable.

The central authority would supervise the maintenance and improvement of the system, oversee its finances, supervise citywide signal carriage, establish interconnection policy, and collect fees from vendors using citywide and local channels in order to support central and local cable-related activities (for example, programming, management, elections, and so forth). Strict rules would be laid down to ensure the privacy of privileged communications. Users might or might not be required to pay for cable service; since a local community would negotiate with vendors, trading access to the community for services desired, it might realize sufficient revenues to offset operating costs—or it could be forced to bill its user population to meet those costs when particularly desirable services had been allowed access at no charge. The central authority would judge the necessity of subsidizing vital services for communities unlikely to attract vendors.

However, it is one thing to state the need for a proposed system, and quite another to bring it into existence. Georg Lukács observes that as a first step:

> Every "theoretical" tendency or clash of views must immediately develop an organizational arm if it is to rise above the level of pure theory or abstract opinion, that is to say, if it really intends to point the way to its own fulfillment in practice. . . . [An] analysis that would see an organized action in terms of the lessons it contained for the future, as an answer to the question "what shall we do?" sees the problem in terms of organization.[28]

The advocate having become, of necessity, an organizer, his or her responsibility is as much to the community as to personal visions (unrealizable without community help). This responsibility requires knowledge of the new arena. According to David Morris and Karl Hess,

> The first task of an organizer is . . . to discover what the neighborhood is, and this is done not by proclaiming programs, but by listening. "Actually," notes Alinsky, "you do more organizing with your ears than with your tongue." Listening helps to discover not only the customs of the neighborhood, but what its residents believe its problems to be, who the natural leaders are, and what the history of struggle and organization has been in that area.[29]

Then,

> as the first institutions and communications systems develop, a neighborhood begins to have hope, to contemplate the possibilities. At this stage, maybe a small number of area residents will begin to be involved along with the counter-culture community in developing new values and meeting old needs. . . . It is at this stage . . . that the next, and more long-lasting stage can be initiated, that of controlling increasing parts of the neighborhood's commerce, and establishing new productive facilities to meet the needs of residents.[30]

This process is, at least in its early stages, arduous and demanding. The organizer must be everywhere, responding to community opinion as much as to the requirements of the developing institutions—in the present case, the public cable system. The role of organizer requires great fluidity and originality; there are no scripts to follow, especially when dealing with issues that may seem to the community ephemeral, utopian, or simply irrelevant.

Morris and Hess concentrate upon the development of the neighborhood as a first priority in the construction of a new America, but the advocate of a municipal cable system will have to coordinate political activity throughout the city. Allies will be needed in city hall to remove obstacles to the organizer's own efforts. However, responsibility must not devolve upon the city administration to work out the strategies necessary to develop the municipal cable system. The people must participate in planning and implementation through their own political activity and through city-sponsored programs, using such tools as trade-off games, Delphi techniques, and other effective planning techniques. This participation is crucial to the future success of the system: the clients of the system must have a commitment to make it work *after* the physical plant is in place. Morris and Hess rhetorically ask:

> Will the people of the neighborhood participate in the decisions that affect their lives—no matter how time-consuming, no matter how troublesome? Will they be full-time citizens or just now-and-then voters?
>
> When the people, or at least many people, in a neighborhood finally decide that they want to participate in politics and not just delegate power to politicians, then community politics begins, local liberty becomes a possibility, and the debates over neighborhood government sensibly begin.[31]

Substitute "community" for "neighborhood," and "communications" for "politics," and the argument holds as well for cable as for any other local institution:

> Communication is the key to all of this. Without the kind of communications which permit the average citizen to interrelate with other citizens, trade becomes the province of business people, politics is governed by representatives, and intercommunication becomes much like internationalism; that is, relations between sovereign states, between diplomats who are emissaries of duly elected leaders.[32]

A good example of a local community regulating its own affairs at the participatory level of government is the town of Tapiola, in Finland. Tapiola is not an entirely spontaneous happening: it is the reasoned product of Asuntosaatio, a nonprofit new town-development corporation underwritten by the Finnish Housing Foundation and a half-dozen labor and trade associations.[33] Its council of 16 members includes economists, engineers, architects, and other "technical specialists," as well as representatives of the member associations. In their employ are three departments—planning, technical, and administrative—which develop and service the new towns sponsored by Asuntosaatio.

In Tapiola (built in the last decade), nearly 90 percent of all dwellings are owned by their occupants. In practice, Tapiola is something like a supercondominium:

> A joint-stock "housing company" is formed for groups of buildings of suitable size before construction is started. . . . The company owns the building comprising a group, and the resident owns a certain number of shares entitling him to hold a specified flat or dwelling unit. The company is managed by a board that is appointed by the occupants at an annual company meeting.[34]

But the residents are more than self-interested homesteaders on a vast corporate farm. They participate in the production of vital services with an intensity unseen in the United States:

> All joint-stock housing companies in Tapiola are stockholders in the so-called "supply companies," . . . [which], in turn, own the heating plant, certain business premises and buildings, central garages, and a building in which the present cinema is located. . . . The board of directors of [the major supply company] consists entirely of representatives of the housing companies in Tapiola.[35]

The Tapiola philosophy encourages ecological management, as well:

> The supply companies in Tapiola formed an independent service company . . . which is in charge of landscaping and grounds. This company also maintains flower beds. It is, as well, responsible for the collection and incineration of waste. . . . [Another service company] is responsible for the distribution of electric current, which is a by-product of the central heating plant [which runs on the incinerating waste].[36]

Most important to the concept of Tapiola, however, is its totally democratic mode of operation. In its initial stages, the professional departments were guided by the Asuntosaatio council, in cooperation with local residents, who participated in architectural planning and implementation. But once the city is completed in all respects, "its administration will be completely in the hands of residents experienced in its management."[37] In the meantime, should things not fit resident needs,

> There are three ways that a resident of Tapiola can propose an idea or challenge a decision of Asuntosaatio:
> 1. A personal call to the director; this is frequently done.
> 2. A letter to the editor of the town's newspaper.
> 3. A sharehold in a housing company (of which all residents are members) can insist that the company's board challenge an action or decision.[38]

The near complete democracy of Tapiola, combined with its public enterprise economy, is a remarkable example of enlightened public planning in an enlightened country. The advocate of public cable in the American city—whether video enthusiast, scientist, municipal officer, or private citizen—will find, as community organizer, that there are no urban "blank slates" upon which grand visions can be inscribed. Established custom, institutionalized ignorance, and active opposition from threatened corporate entities and their regulatory allies will confound all strategic plans and call for appropriate tactics on the part of the organizer. Immersion in local politics may become necessary, as may unexpected alliances with community leaders previously thought to be unconcerned with the cable issue.

But the potential public rewards for perseverance, critical analysis, and success will continue to mobilize Americans throughout the nation in pursuit of public control of communication media in general and broadband communications in particular. The continuity of this effort suggests the appearance of a new brand of twentieth-century community, unafraid of modernity as it strives to reclaim an American legacy of local independence and intercommunity cooperation.

AN "IDEAL" SYSTEM

Assuming that a city and its citizens could create a strategy for development of a complex urban telecommunications system, and given a social, political, and economic environment favorable to that development, what form might the "ideal" cable communications system take?

It would be universal in distribution. Utilizing existing city facilities—conduits, poles, microwave links—and newly installed plant, the system would have major trunks running throughout the city. At selected nodes on the network, data-processing equipment would switch digital data (from video and audio sources, as well as from computers) among and within local districts.

These districts would have considerable autonomy in establishing the mix of communications entering their districts. (District boundaries would be based on planning districts, ethnic subcommunities, or other socially determined divisions; they would not be based on traditional, irrational, and too easily exploited political entities—such as councilmanic districts.) Communities would elect "cable boards" to sponsor local, intradistrict programming; to delegate representatives to sit on an interdistrict cable board; and to evaluate applications from vendors wishing to market services within the districts.

They would have this power, since the entire system would be franchised to the city; *channel leases* would be required of all local and interdistrict service vendors and programmers. Leases would be negotiable as to duration and costs. However, for every district channel leased, a tariff would be imposed by the city. A portion of tariffed revenues would be returned to the districts on an equitable basis; the remainder would be used to repay municipal debts incurred during system construction—and then to offset operating costs. Since local groups and entrepreneurs would only have to advance capital for leasing a channel in one or more districts—and not for construction of the entire plant—local programming could increase in quantity and diversity.

With channel tariffs paying the greater proportion of system costs, subscriber fees would be kept at a reasonable level. Some classes of users—the elderly, the poor, apartment dwellers—would be offered subsidies. (At some point, when the city believed it had created a sufficient reserve to cover the costs of upgrading the system on a regular basis, fees might be reduced entirely.) Subscribers could select the precise channels (in addition to municipal, educational, and other public originations) for which they were willing to be charged by the channel vendor. Some service channels might cost nothing to obtain, even though they originated with commercial firms; banks might offer free access to attract customers or reduce their own operating costs, as might supermarkets and other merchants.

Billing would be handled by the city, with its comprehensive access to all records of signal transmission and exchange. To prevent a glut of commercial services, however, multiplexing of signals and sharing of tariffed channels by associations of vendors would be encouraged. Final determination on the issuance of all leases would reside with an interdistrict cable board, in accordance with the recommendations of the various local cable boards, which would be legally binding.

Since the system could only support a finite number of signals, some channels would be reserved for public uses. These would be delegated to the various public communities of interest—educational institutions, health delivery services, municipal information agencies, and citizen organizations. Except for simple digital signaling, it is unlikely that homes or offices would produce programming. Production centers would be established by government within local districts to promote local origination: these might be in libraries or other public buildings, providing a welcome opportunity for face-to-face interaction among citizens involved in production. Some of these channels reserved for public use would be interdistrict and citywide; others would be reserved within districts for local uses.

The system would not necessarily compete with existing technologies (for example, the telephone system). It would, instead, concentrate on the transmission of interactive, digital signals, complementing analog audio (telephones) and video (television) transmission offered by the conventional media. The cable communications system would be compatible with other digital transmission systems, including satellite networks and specialized common carriers requiring local distribution facilities.

As companies, government agencies, and third-sector groups (including citizen organizations) desired to establish their own loops—to connect decentralized "work centers" or citizen meeting places, for example—the city would estimate the cost of adding additional capacity to the system and charge the cost directly back to the users, possibly on a lease-back arrangement, causing ownership of the channel to eventually revert to the user.

This flexibility would be reflected throughout the system design. The interdistrict cable board, in consultation with its support staff of city personnel, would decide on dates for the retirement of obsolescent equipment (after an accelerated depreciation period). If a radically new technology should become commercially available—such as the much-heralded but as yet still experimental optical fibers—then the board would direct the city to begin implementing the new technology. (It is likely that optical fibers will be compatible with existing technologies, so that adoption of the new technology would be commercially feasible; the city would not have a difficult time negotiating with a supplier of optical fibers to replace existing coaxial cable.)

(However, it is extremely *un*likely that optical fibers or any other revolutionary communication technology will become a commercial possibility in the 15 to 20 years before the proposed cable system is fully amortized.)

The tariffs on channel rentals would return a large portion of vendor capital and subscriber fees to the city, prompting economic recovery at a pace inconceivable today. Cooperative enterprises would also become a real urban possibility, since organizers could obtain a channel at a nominal cost for communicating among the collective group. New social communities, unrelated to geographical location, would come into being with the discovery of interdistrict or even citywide common interests.

The system would not only afford citizens ready access to each other, their work places, and their elected officials, but it would make possible such innovations in city management as forums and referendums within and among local districts. Savings in costs of data transmission (including reduced transportation costs) would be enormous; the city would become attractive to enterprises seeking the benefits of efficient information accessing and distributing systems for themselves.

However, as important as the fiscal worth of the system might be, the prime virtue of the public cable communications system would be its ability to transmute the usual oppressive social environment of the city: it would provide an infrastructure for the "direct, constant interaction between people, individually and in community," called for by Serge Chermayeff and Tzonis, and enliven "public concourse, the collective interaction, for the structuring of man-made environments of variety."[39]

Would the system be immune to exploitation by ensconced corrupters? This would depend on the vigilance of the people. By making the technical operations of the system wholly automatic—that is, to have the system function as an advanced, interactive common carrier—the opportunities for manipulative abuse would be lessened. With access fairly distributed and ensured by democratic participation in local and citywide decision making, information resources for the masses would be greatly increased. But would the citizens be able to use them to preserve democratic control over the system?

The answers to this and other questions must await the development of the first truly interactive city system. The time may not be far off: with the cities of Milwaukee, Del Mar (California), and the Orange County (California) communities already planning public systems very much like that just described, understanding will come as the first signals—the first communications—begin to flow.

NOTES

1. Murray S. Stedman, Jr., *Urban Politics* (Cambridge, Mass.: Winthrop, 1972), p. 302.

2. Mel Scott, *American City Planning Since 1890* (Berkeley: University of California Press, 1969), p. 650.

3. The Urban Institute, *The Struggle to Bring Technology to the Cities* (Washington, D.C.: The Urban Institute, 1971), p. 40.

4. Kas Kalba, "Telecommunications for Future Human Settlements: A Planning Framework for Minnesota Experimental City," *Ekistics* 211 (June 1973): 332.

5. Ibid.

6. Ibid., p. 336.

7. Martin Kuenzlen, *Playing Urban Games* (Boston: i press, 1972), p. 98.

8. Harold Sackman, *Mass Information Utilities and Social Excellence* (Princeton, N.J.: Auerbach, 1971), p. 242.

9. Erasmus H. Kloman, "Public Participation in Technology Assessment," *Public Administration Review* 34, no. 1 (January-February 1974): 58.

10. Daniel D. Roman, "Technology Assessment: Perspective from the Managerial Position," *Public Administration Review* 33, no. 5 (September-October 1973): 396–97.

11. Ibid., pp. 398–99.

12. Robert A. Dahl, "Governing the Giant Corporation," in Ralph Nader and Mark Green, eds., *Corporate Power in America* (New York: Grossman, 1973), p. 17.

13. Kuenzlen, *Playing Urban Games*, p. 8.

14. Alexander Tzonis, *Towards a Non-Oppressive Environment* (Boston: i press, 1972), p. 110.

15. Ibid.

16. Norman I. Fainstein and Susan F. Fainstein, *Urban Political Movements: The Search for Power and Minority Groups in American Cities* (Englewood Cliffs, N.J.: Prentice-Hall, 1974), pp. 185-224.

17. Franklin P. Huddle, "The Social Function of Technology Assessment," in *Technology Assessment*, ed. Raphael G. Kasper (New York: Praeger, 1972), p. 163.

18. Gerald D. Suttles, *The Social Construction of Community* (Chicago: University of Chicago Press, 1972), pp. 58–61.

19. Edwin B. Parker, *On-Line Polling and Voting*, mimeographed (Stanford, Calif.: Stanford University, September 1971).

20. Seymour B. Sarason, "Community Psychology, Networks, and Mr. Everyman," *American Psychologist* 31, no. 5 (May 1976): 325.

21. Ibid.

22. Laurence I. Press, "Arguments for a Moratorium on the Construction of a Community Information Utility," *Communications of the ACM* 17, no. 12 (December 1974): 678.

23. Hans Magnus Enzensberger, "Constituents of a Theory of the Media," in *Sociology of Mass Communication*, ed. D. McQuail (Baltimore: Penguin, 1972), pp. 114–15.

24. Jacqueline Scherer, *Contemporary Community* (London: Tavistock, 1972), p. 117.

25. Royce Hanson, "Opportunities for Innovative Local Government," in *The Contemporary New Communities Movement in the United States*, ed. G. Golany and D. Walden (Urbana: University of Illinois Press, 1974), p. 79.

26. Ibid.

27. Los Angeles, Study Team on Franchise Districts, "Interim Report on Franchise Districts," in Citizens Task Force on Cable, *Summary of Interim Reports of Task Forces* (Los Angeles: Los Angeles City Department of Public Utilities and Transportation, 1976).

28. Georg Lukács, *History and Class Consciousness* (Cambridge, Mass.: MIT Press, 1971), p. 299.

29. David Morris and Karl Hess, *Neighborhood Power.* Boston: Beacon, 1975), p. 40.

30. Ibid., p. 37.

31. Ibid., p. 104.

32. Ibid., p. 155.

33. Heikki von Hertzen and Paul D. Spreiregen, *Building a New Town* (Cambridge, Mass.: MIT Press, 1973), p. 164.

34. Ibid., p. 170.

35. Ibid.

36. Ibid., pp. 170–71.

37. Ibid., p. 171.

38. Ibid., p. 169.

39. Serge Chermayeff and Alexander Tzonis, *Shape of Community* (Baltimore: Penguin, 1971), pp. 191–92.

CHAPTER

9

SUMMARY

Many forces are converging to make municipally owned and operated cable communications systems attractive to innovative urban officials and citizens alike—the worsening social and economic crises of the cities, the growing awareness of citizens that they can have real power in the creation of their environment, and the failure of commercial cable television to serve the needs of cities and citizens alike. It is unfortunate, in a sense, that no commercial operation has been successfully integrated with a large working urban community: with no "real world" models to look to, only conjecture about the socially progressive capabilities of cable communication is possible. But the consensus among most communications researchers, policy planners, and the few citizens who have had positive experiences with cable television is that there is a potential for radical change in the structure and workings of the city—even to the extent of creating communities able to save the settlement from imminent decay—realizable with the aid of cable technology.

There are a few relatively small-scale municipal cable television systems in operation (as well as in the planning or evaluation state):

Municipal Systems Presently in Operation[1]	City Systems Planned or under Evaluation
Opp, Alabama	Berkeley, California
Bella Vista, Arkansas	Del Mar, California
San Bruno, California	Fountain Valley, California*
Valparaiso, Florida	Huntington Beach, California*
Frankfort, Kentucky	Newport Beach, California*
Walkertown Station, Kentucky	Palo Alto, California
Crystal Falls, Michigan	Westminster, California*

Norway, Michigan
Jackson, Minnesota
Newburg, Missouri
Greeneville, North Carolina
Cascade Locks, Oregon
Blossburg, Pennsylvania
Pitcairn, Pennsylvania
Rouseville, Pennsylvania
Sumas, Washington
Rivesville, West Virginia
Boaz, Wisconsin
Steuben, Wisconsin
Woodman, Wisconsin

Santa Barbara, California
Jacksonville, Florida
Brookline, Massachusetts
Detroit, Michigan
Madison, Wisconsin
Milwaukee, Wisconsin

*Member, Public Cable Television Authority, Orange County, California.

The extant systems are by no means oriented toward the involvement of citizens in their operations. They are, however, efficient providers of conventional cable television service—and there is hope that the future will see more citizen participation in their programming and management processes.

But the much larger systems that are planned *do* make provision for considerable citizen involvement (see above). These larger systems—citywide or regional cable networks—will cost tens, and even hundreds, of millions of dollars and will test the contention that investment in cable communications can better the city, not only as a marketplace but as a place to live as well.

Harold Horn, of the Cable Television Information Center, has expressed the belief that there are no clear differences between public and private ownership in terms of system management; there is only the question of "which ownership form do you trust to operate the system best—public ownership motivated by the public interest or private ownership motivated by profit (within a regulatory framework)?"[2] The city's answer to this question will be decisive—for, as Ted Ledbetter contends, "ownership is important . . . because the ownership of the system is the only way to guarantee open access."[3] The history of broadcasting and other mass media is one of private ownership consistently being used to deny any but the most limited media access to the general public; unfortunately, it is these media that own the majority of cable television systems.

Access advocates have found little sympathy at the Federal Communications Commission, which has systematically "unburdened" the cable operators of one after another public service commitment. The most recent action of the FCC, voiding the 1977 deadline for cable reconstruction (including upgrading for two-way capability) for older systems, could prevent two-way capabilities from developing in the near future. The decisions of the FCC have been well coordinated with the plans of cable operators to offer more pay-services—pay-movies and pay-sports predominating—to affluent cabled suburbanites, while ignoring the central cities, where most Americans live.

The cities continue to decline as livable environments even as they expend immense amounts of capital and labor accommodating exploitive, largely non-local corporations. The cities are, today, finally faced with a problem that is "no longer 'simply' that of physical rehabilitation and expenditure of billions of dollars, but of formulating and of translating into reality reasonably clear and acceptable ideas of what kinds of cities we want, and how we can make them more attractive and humanly meaningful to the nation's urbanized population."[4]

During the Delos Four conference, a resolution was approved that called attention to the central role communications play in the sustenance of community and purposeful human activity:

> Much of our communication is a one-way system. There is no means of setting up a dialogue. In former times man received all information about his family, most about his neighborhood, less about his city, and very little from the outside world. Now we receive world news in an instant, but have lost contact with those who live close to us. . . . We need to strengthen man's ability to process and pass on information to his family and small group, but this means there must be opportunities for dialogue.[5]

Urban cable communications could enable this dialogue to grow in the city, among individuals and among groups; in the process of implementing such a system, citizens could gain the experience and confidence necessary to manage the civic machinery in a democratic manner, without domination by small but better-informed elites. In order for urban societies to effectively use these real-time information systems, Harold Sackman strongly advises that "a new level of participant democracy is needed well beyond anything that has been attempted so far. . . . The alternative is the eclipse of the public by a new technological aristocracy."[6]

A Government Committee on Communication Policy, established by the Parliament in Finland, concluded that in Finland at least, "cable communications should be developed along with a comprehensive communications policy and general social planning. Only on this basis can cable communication be rationally developed in relation to society's needs."[7] The committee was convinced, in light of that conclusion, "that social equality and freedom of speech can best be achieved in cable communications when both private and mass communication take place free of strong economic interests, with publicly-owned media."[8]

(As crucial as public ownership of the cable medium would be a strong public role in planning for, and management of, the public cable system. A major problem for public organizations and institutions operated independently of citizen-clients is an inherent lack of feedback from these clients; unlike private businesses, where there is no easily identifiable indicator, such as "profits," by which these agencies can measure their effectiveness.[9])

(In recommending public ownership of the cable system proposed for Detroit, the Cable TV Study Committee there suggested that management responsibilities could be delegated to cable experts well insulated from the traditional politics of the city.[10] The motives of this Committee were noble but naive. City politics "as usual" cannot be undone by condemnation. It would be more honest and effectual to directly address questions of urban politics, before, during, and after cable system planning and implementation. This is especially true in ravaged Detroit, run by its automotive baronies; the naivete of the Committee in proposing a nonpolitical public system may account, in part, for the system remaining unbuilt five years after the Committee's report.)

Community ownership, often advocated as a more desirable form of "public" ownership than ownership by the city (which *is*, at present, dominated by elites), is a concept that has proven unfeasible. Community ownership schemes have not been able to generate more than verbal enthusiasm. Those community-owned systems that are in operation usually owe their success to the political ties and financial aspirations of local entrepreneurs, who are hardly reflective of the socioeconomic character of their communities.[11]

The *Yale Law Journal*, apprehensive of the loss of autonomy a public system might suffer during a political crisis, has suggested the creation of a separate public corporation (as in Frankfort, Kentucky) to manage the system.[12] (See Appendix.) This is a plausible solution to the problem of potential interference by exploitive interests in the operations of a cable system managed by a government dependent on those same interests. It would require, however, that the appointment of directors to the corporation be made in some other manner than that currently practiced—the selection of prominent members of the commercial class who set policy "in the public interest." The public corporation should be as accountable to the people it serves as a publicly owned system operated by the city. Inasmuch as the public corporation will only reflect the democratic or oppressive nature of the municipal government, it might be more productive for citizens to concentrate on democratizing local government rather than dreaming up new schemes for isolating the cable system from "politics."

(These activists should take note of the growing involvement of existing operators—and their opponents, the common carriers and the broadcasters—in local politics. For example, Theta Cable of California, while sufficiently ailing to request a 42 percent subscriber rate increase in 1974, still found over $10,000 in discretionary cash for disbursement to members of the Los Angeles City Council.)

Louis Schwartz and Robert Woods, writing for the PubliCable public interest group, have suggested allying the cable system with some other municipal agency—possibly the municipal power utility:

> . . . the community character of public power systems makes them uniquely qualified to bring to fruition many of CATV's potentialities for local public service. With its sense of responsiveness to the local community, public power can assist cable in reacting to and meeting community problems and concerns. . . .
>
> There have been complaints . . . that municipal or non-profit groups lack the initiative and innovative spirit to bring vitality to CATV. The record of many public power systems establishes clearly that no such difficulties would inhere in any public power entry into the cable field. Indeed, the new challenge of cable might infuse settled public power systems with a new sense of vitality.[13]

The admiration of Schwartz and Woods of public power agencies seems well founded in light of events in California, where "taking arms against a sea of troublesome power-rate increases, . . . publicly-owned utilities are launching a fleet of power-generation enterprises ranging from a geothermal steam venture to coal-fired plants and public-private nuclear-power projects."[14] How accessible the direction of these well-established bureaucracies would be to local citizens is problematic, though the notion of such an alliance is an appealing one (assuming accessibility could be guaranteed).

There have been many exciting experiments performed with two-way cable communications, like the unexpectedly successful "telepsychiatry" clinic run by Boston's Massachusetts General Hospital at Logan Airport, three miles away,[15] or the computer-coordinated electric power systems now proliferating throughout the nation.[16] Unfortunately, the cable industry has been either unwilling or unable to demonstrate two-way services on any reasonable scale in a populated urban area, despite the readily availability of technical capability. Advocating increased business data use of cable television services, especially by nationwide teleprocessors who are in competition with the telephone companies, Joseph Stern of Goldmark Communications claims that

> many of the "blue sky" services were adaptations or versions of existing business communications. High-speed data transmission, high-speed facsimile, interaction between intelligent terminals—all are services presently in use by the business community. . . .
>
> The business community has the need for transmission of vital information and is ready to pay for cost effective methods of providing this information flow and interaction.[17]

Like the business community, the *citizens* of the city have a need for transmission of vital information—what is urgently required is an operating system to allow the need to be popularly conceptualized and solutions actualized! Without such an initiative, predicts Sackman, and

> with unrestricted laissez-faire, mass computer services [probably on cable channels] would follow the path of least resistance and greatest profits, probably along the lines of pay-TV entertainment on demand. With well-formulated and widely supported public policy and adequate government support, universal lifelong education and home voting systems could be available at the outset and could continually dominate the long-term evolution of computer services. . . . many would like to see free educational, health, and welfare services, but recoil at the cost, and resign themselves to the vagaries of the marketplace as the more "realistic" approach. But an order of magnitude increase in the use of computers can lead to startling new economic considerations.[18]

There is a definite problem in reaching some threshhold of cable "penetration" that would make many services now considered "blue sky" economically feasible. A new sort of advocacy economics is required to evaluate potential services and to determine the present costs of *not* having them available.[19]

George Gerbner would assign this analysis to the communications specialists and scholars, who "have a special responsibility for calling a halt to the policy of drifting with the tide and to the process of deciding by default."[20] They can give aid, but they are in no better position to evaluate the impact of urban communications than the citizen upon whom the impact falls. The final determination for or against advanced urban telecommunications must come from the urban masses themselves.

If this causes some officials, and even some citizens, to stop and take a second look at cable, as Boston's Mayor Kevin White has done, well and good; patience is a virtue when the only alternative is hasty franchise activity by the city, resulting in inferior service agreements lasting 15 years or longer. But, eventually, some action must be taken. When it is, it should be in the interests of the people who live in the city and who work there, and in whose hands rests the power to restore the city's vitality. The Detroit Cable TV Study Committee, in a particularly eloquent passage, explained why that city is moving slowly—not to forestall making decisions but to ensure that the right ones are made:

> If the insistence upon guarantees of public protection delays the advent of cable television in Detroit, the delay must be endured. The guarantees are essential, and they must be incorporated into a master plan *before* any system is allowed to begin. To do otherwise is to repeat the errors of the past: the errors of first allowing trees to be felled without restrictions, then wondering why Michigan has no forest; the errors of first condemning buildings, then worrying about relocating the people in them; the errors of first building freeways, then dealing with the exodus of the more mobile families from the city.[21]

It has been argued that urban populations should make a public investment in cable communications, to be planned and managed by the people of the cities. It has been claimed that such an effort could have useful effects on the organization of the city, causing the great majority of uninvolved and exploited urban dwellers to realize the power that resides in their skills and labor. It has been maintained that this is all possible—and yet, it is impossible to point to even one working municipal system that has fulfilled all of those promises. Above all, it has claimed that such an undertaking, together with the communications infrastructure it would create, would radically alter the city as a social organism—causing it to resemble more closely the democratic and independent city-states of the early Renaissance, the birthplace of modern humanism.

Clearly there is an element of utopianism in this prediction—and as Alexander Tzonis has so critically observed, a utopia "stands somewhere between being a joke, a mild violation and a crime, a forceful breach."[22] There is danger to this: "At the very worst, 'alternatives' are incorporated within the existing *status quo* to create an illusion of openness, flexibility, and its desire for improvement where there is in fact only authoritarianism and despair."[23] This has not been the purpose of the present discussion. In admonishing the advocates of municipal ownership to consider the role of the people in implementing cable communications, it has been of the utmost importance to repeat, over and over again, that cable is only a technology—and that the only permanent solutions to the city's ills lie in a drastic reorganization of its internal *social* relationships—which can be *aided* by cable communications properly employed.

The concept of the communication infrastructure—a focus for widespread organizing throughout the city—is being put to the test in locales around the world. The impending development of large-scale cable communications systems sill make possible a demonstration of communication technology used with progressive intent—with a view toward the city as an urban environment in which life can be a direct, comprehensible, and worthwhile experience.

NOTES

1. The list was compiled from information obtained from Cable Television Information Center, Washington, D.C., and PubliCable, Inc., Washington, D.C.

2. Harold E. Horn, of Cable Television Information Center, to Tim Arnold of Columbia, Mo., private communication, February 5, 1973.

3. Ted Ledbetter, "Broadband Communications in an Urban Setting," in *The Wired City of Los Angeles*, ed. Robert Warren (Los Angeles: Center for Urban Affairs, University of Southern California, 1972), p. 7.

4. Robin Marris, "Business, Economics and Society," in *Social Innovation in the City*, ed. Richard S. Rosenbloom and Robin Marris (Cambridge, Mass.: Harvard University Press, 1969), p. 13.

5. Delos Four, "Need for More Balance in the Flow of Communications," in *Human Identity in the Urban Environment*, ed. Gwen Bell and Jacqueline Tyrwhitt (Harmondsworth, U.K.: Penguin, 1972), p. 517.

6. Harold Sackman, *Mass Information Utilities and Social Excellence* (Princeton, N.J.: Auerbach, 1971), p. 252.

7. Finland, Government Committee on Communication Policy, *Investigations and Recommendations Concerning Developments in Communication Technology (Abridged)* (Helsinki: Finnish Council of State, 1974), p. 40.

8. Ibid., p. 41.

9. Marris, "Business, Economics and Society," p. 19.

10. City of Detroit, Cable T.V. Study Committee, *Cable Television in Detroit* (Detroit: City of Detroit, 1972), p. 20.

11. Barry Lemieux, "Ownership Patterns and Alternatives," in *The Wired City*, ed. Warren, p. 75.

12. "Toward Community Ownership of Cable Television," *Yale Law Journal* 83, no. 8 (July 1974): 1724.

13. Louis Schwartz and Robert A. Woods, "A Marriage Proposal: Cable Television and Local Public Power," *Public Power* (January-February 1972): 1.

14. Steve La Rue, "The Small Cities' Threat to the Giant Utilities," *California Journal* 6, no. 9 (September 1975): 318.

15. "Turning on Telepsychiatry," *Human Behavior* 3, no. 5 (May 1974): 30.

16. Hans Glavitsch, "Computer Control of Electric-Power Systems," *Scientific American* 231, no. 12 (January 1975): 34–44.

17. Joseph L. Stern, "Business Data on CATV Possible with Wideband Analog Cable Systems," *Communication News* 12, no. 1 (January 1974): 30.

18. Harold Sackman, "Computers and Human Problems," in *Computers and the Problems of Society*, ed. Harold Sackman and H. Borko (Montvale, N.J.: American Federation of Information Processing Societies [AFIPS] Press, 1972), p. 25.

19. William H. Melody, "The Role of Advocacy in Public Policy Planning," in *Communications Technology and Social Policy*, ed. George Gerbner, Larry P. Gross, and William H. Melody (New York: John Wiley & Sons, 1973), p. 179.

20. George Gerbner, "Preface," in *Communications Technology*, ed. Gerbner, Gross, and Melody, p. x.

21. City of Detroit, Cable T.V. Study Committee, *Cable Television in Detroit*, p. 20.

22. Alexander Tzonis, *Towards a Non-Oppressive Environment* (Boston: i press, 1972), p. 107.

23. Ibid.

APPENDIX
THREE OPERATING PUBLIC CABLE TELEVISION SYSTEMS

Of the over 20 municipal cable systems in operation or planned, the majority serve relatively small communites. These are, for the most part, operated as municipal utilities—much like municipal water or electrical power utilities. As publicly owned systems have begun to serve large communities, however, the methods of operation—including financing, marketing, and general management—are becoming more complex and sophisticated.

The three largest municipal cable systems—operating systems in San Bruno, California, and Frankfort, Kentucky, and a system planned for the Public Cable Television Authority of North Orange County, California—demonstrate this diversity well. While the two large operating systems have been successful in providing high-quality services economically (as have the smaller systems, to varying degrees), neither operates as a true "communication infrastructure." Community involvement is, at best, limited to electoral influence; consequently, while the public systems have displayed a vastly superior capability to provide conventional cable television services when compared with most private systems, they have not demonstrably contributed to the solution of urban problems through increased public dialogue and organization.

Undoubtedly, this reflects the urban environments in which these systems came into being: both San Bruno and Frankfort are small, homogeneous suburbs, without citywide problems of great consequence awaiting solution. Also, the technological sophistication of these systems, while high, has not been used for two-way communications of any kind; channels are not being used for nonconventional, informational services. However, as San Bruno's director of Public Works, Leo Van Dusen, has written, "We have not had one request as yet for any access programming."

The cable communications system planned for the participating cities of the Public Cable Television Authority (PCTA) is more ambitious in conception. The system will serve nearly 200,000 citizens, compared with San Bruno's population of 40,000; it is envisaged as being completely two-way, data-capable, and largely accessible for local programming. However, citizen participation in the planning of the system has been extremely limited, and it will be interesting to see how the actual system design fits local needs and wants.

A cable communications system designed with full citizen involvement in planning and operation might or might not resemble these three systems in physical dimensions or technological capability.

The following documents describe, in the words of those responsible for their management, the origin of the three systems representing alternative forms of publicly owned cable systems. (A brief commentary precedes each collection of materials.) Whether these systems will evolve into truly integrated urban communication networks is problematic; the PCTA plan, of course, is designed to perform in the role of infrastructure, while the other systems only provide conventional cable services.

The success of these public systems contradicts the disparagements of private enterprise proponents: not only do these systems (those that are built) operate efficiently but they generate earnings for the cities while producing savings for their subscribers. However, they should not be viewed as necessary models for cities contemplating municipal cable communications systems. Each city preparing for municipal cable communications will discover its own unique resources and needs, which will be reflected in the cable communications system it finally operates. As these three public operators have shown (along with a score of smaller cities), publicly owned cable systems are feasible and desirable—and, ultimately, realizable.

SAN BRUNO, CALIFORNIA

A completely city-run system, San Bruno Cable Television is the largest municipal cable television system currently in operation in the United States—or the world. San Bruno (population 37,500) is a suburb of San Francisco, lying a few miles south of it; it is essentially a "bedroom" community, though it does have a fair number of commercial establishments within its borders.

San Bruno, like many other cities operating cable television systems, has a long history of municipal provision of vital services; the cable system was initially funded out of Water Department revenues (though the debt has since been repaid). Revenues from the cable system are merged with the Water Department's revenues; this provides a continuity of cash flow for the cable system, with all borrowed funds being repaid within the fiscal year. According to Director of Public Works Leo Van Dusen, in a letter dated April 22, 1974, the question of incorporation of funds "is a real old one . . . that . . . presupposes the CATV system will not be able to support itself. . . . This of course is not true because each of our CATV budgets . . . is balanced with the fiscal year's CATV receipts."

The City Council, without urging from the FCC or other external regulators (the San Bruno system is "grandfathered," exempting it from compliance with FCC performance regulations), has determined that the system must provide three hours of local programming each night. Along with city departmental programming, courses in career education were produced with a local junior college, and the "local Posy Parade . . . was received so well we were requested to replay it a total of seven times," according to Van Dusen. The system has recently purchased additional production equipment, and plans have been made to double the production budget allocations for salaries and materials. Van

Dusen is currently investigating the possibility of exploiting the two-way capabilities of the system.

Providing nearly 20 channels of rebroadcast and origination programming, the system has achieved outstanding levels of penetration in every district of the city; this was not foreseen in a research report prepared by the Stanford Research Institute, which forecast a moderate success for a municipally-owned cable system in San Bruno, based upon high penetration in the hilliest quarter of the city. The fifth and last cable district of the city was only recently completed, with service now extended to 100 percent of the city; in the four previously cabled districts, penetration is an encouraging 50 percent, with 4,045 residential connections and a remarkable 3,136 commercial contract connections (large apartment structures and business establishments) at the start of 1975. Rates are an exceptionally low $15.00 for installation and $4.75 per month. With yearly profits approaching $150,000 (for reinvestment in the system), a rate increase still seems some months away.

San Bruno undertook its planning in a very formal manner, without assistance from outside agencies. In so doing, it has been subjected to substantial criticism by the NCTA and other proponents of the private enterprise model for cable development. In countering these attacks, the city has adopted an almost evangelical posture, seeking support for the concept of municipal ownership among American cities.

The city, upon final completion of the system, has begun the necessary procedural activities to secure certification from the FCC; achievement of certification will mark a major victory for the advocates of public ownership of the cable medium.

Will the San Bruno system operate as a true "communications infrastructure" to the benefit of a community, with needs to be met through citizen action? As a suburb of limited size, communication in San Bruno seems adequately diverse and pervasive to eliminate the need for a cable network superimposed upon more conventional channels of communication. Its homogenous population is already highly involved in civic affairs: typically, the city polled its citizens before embarking on the development of the system. Nevertheless, as Van Dusen admitted in a letter of July 1, 1975, ". . . it is our opinion that we have the support of the citizens of San Bruno who feel that we are doing a good job in this utility as we are in all our other utilities. We have had no comments or questions regarding our utilities, including cable television."

But, "There have been some requests for expanding our programming."

The San Bruno system operates efficiently. It certainly exceeds the expectations of all communications planning experts. But does it adequately take into account the needs of urban citizens, who intuitively perceive a greater role for themselves in the operation of the cable utility? How can the city integrate citizens into an organization already institutionalized as part of the city government? These are questions that San Bruno—as well as the rest of the public systems, and those cities contemplating starting public systems—will have to consider more carefully than they have in the past.

Presented to the National League of Cities at San Juan, Puerto Rico

MUNICIPAL OWNERSHIP—A CITY SATISFIED

GERALD D. MINFORD
City Manager
San Bruno, California
December 1, 1973

Controversy surrounding the question of municipal ownership has been answered for the citizens of San Bruno, California. This community of 37,000 finds municipal ownership the best way to provide cable television to its citizens. Early anxieties which are always a companion when large sums of money are invested in a new field which has no precedent for municipal ownership and operation have been allayed. The system is operative, the quality of signal is excellent, and revenues have exceeded expectations.

The City has a 21-channel capacity, push pull, automatic level control amplifiers with capability for future 2-way, and .750 solid aluminum sheath cable; 67 miles of cable have been installed with only 10.7 miles of cable remaining to be installed. The signal level was set to hit each amplifier in line at a minimum level of 13 decibels and padded down to allow a safety factor against future reduction of a signal level due to the aging of the cable. A problem experienced by some private franchisees by over-extending amplifier spacing to reduce cost. I believe the combination of excellent construction, good signal, and first class service are major reasons for our success. We also believe that the lack of that combination is often the reason for some private franchisers' troubles.

This system now serves 6,850 homes and apartments with 1,350 extra set connections. This represents a 57% penetration based upon the FCC formula. We expect penetration to exceed 60%; as the potential of cable television unfolds an even higher penetration will be attained.

Financing of the cable project required an initial investment of $542,000 which was a loan from City reserve funds to be repaid at 7% interest over a period of 7 years. All revenues received above and beyond operating expenses for the first 3 years are being used to complete construction of subsequent phases to extend the trunk lines to the rest of the community and to buy program origination equipment. The City considers interest, a 5% in line tax and a 10% profit as part of operating costs.

The success of the cable system has permitted acceleration of the construction program and will permit repayment of the loan in about 5 years instead of 7 years. After re-payment of the loan the City anticipates an annual profit of $150,000 to $200,000 to its general fund. This can be used to provide a new CATV service, or it can be used for capital improvements, cultural activities, better parks, or to reduce the tax burden. The options belong to the City Council. The return to the City of San Bruno, on a franchise basis using a 5% franchise fee, would only have returned $18,000 per year to the City. This City's goal was to construct and maintain the best possible cable system using the finest equipment and materials available in the field at the time of construction with the ability to grow and adapt to new improvements anticipated in the field and to provide a long term financial benefit to the community.

Our examination of franchise issues in other cities in our county showed a very poor track record, as a general rule, for the private franchisees. Contractual agreements for undergrounding and time limits to complete cable installation were not fulfilled. Cable signal, in some instances, was poor and service on the cable was oftentimes inconsistent. Correction through the franchisee was extremely difficult, and complaints to the City Council to correct the situation too often fell on the deaf ears of the franchisees, who were, oftentimes, busy in other cities in some phase of a new installation.

Dissatisfaction with the experience of many other cities operating under private franchises, and the opportunity for the utilization of profits from cable television to reduce the property tax load on our citizens, led to the investigation of the City constructing, owning and operating its own system.

To investigate the potential market and to determine whether or not the City should seriously consider CATV as a public utility, 10,000 postcards were sent to residents through our water billing office. People were asked if they would use the cable and if they favored municipal ownership. The results of the survey were remarkable; 37% returned cards and indicated that 60% would subscribe and 64% favored a City project. Based on this response, detailed economic studies were made which produced a decision to have a municipally owned and operated cable television system.

San Bruno's entry into cable television shook the cable industry. They launched a vicious, dishonest, unethical attack to discredit the San Bruno CATV system. Every effort was made to make it appear that the leaders of San Bruno placed their community in jeopardy just short of a total financial and political collapse of their local government.

If some of the private companies had entered into community franchises based upon as much research and with as much business acumen as the leaders of the City of San Bruno, they would not have so many losing or marginal operations to use as examples of pending doom for any city that might consider public ownership. San Bruno responded with an information program to tell the truth about the San Bruno experience. The effort to discredit this municipal CATV system has failed. The number of jurisdictions looking at public ownership of CATV systems is increasing at an accelerating pace. Why the massive effort to discourage public ownership? Why is there so much interest in public ownership? Why is the private sector lobbying to take away your right to determine the future of cable television within your community?

Private cable operators have already been successful in obtaining laws which took away your right to negotiate a reasonable franchise fee for the use of public property and rights of way through your city. The right to franchise and the use of public rights of ways throughout the City belongs to the people. The best possible use and return on the use of those rights is a justifiable goal and position for any Council to support. Lobbying is now divided to take away your right to decide between public or private ownership and to eliminate local ability to regulate.

They first tried to scare cities to prevent investigation into public ownership as a viable alternative to private monopolies. The latest "Spectre of Doom" is the image of an octopus-like control of news and programming by government. Hogwash! Existing law and the constant vigilance of the public and journalists across this land prevent business or government from controlling lines of communication. The ability to turn to a different channel, to read a newspaper or magazine prevents thought control. An educated populace cannot be controlled. Our thirst for knowledge and truth is the backbone of democracy. Any effort to slant or control is immediately attacked by journalists and individuals. This new "Spectre of Doom" is a "paper tiger".

It is important that leaders in cities actively oppose any effort at the State or Federal level to take away your right of local self determination of the future of cable television in your community. The future of cable television and its potential is fantastic. It is not just the distribution of entertainment . . . it is a vehicle for dissemination of knowledge and information and services heretofore not generally understood by the general public. In my opinion, cities throughout the country are investigating municipal ownership because the private sector has too often failed to meet its commitments and responsibility to provide the level and quality of service which the public has a right to expect.

Municipal ownership in San Bruno is a financial success and community development of local programming, public access, educational and cultural programs at a level deserved by the people is assured through local control. This does not, however, mean that every city should own and operate its own system. To own or franchise must be a cold, hard business decision. It must be a decision based upon market analysis, community economics, community objectives, and a thorough study of all the alternatives including the private franchise. There are many competent and honest private cable companies who can under a *good* franchise provide CATV service to your community.

DON'T be intimidated by fear tactics from studying public ownership. *DO* oppose any effort to take away your right to do so.

SAN BRUNO MUNICIPAL CATV DEPARTMENT FOURTH ANNUAL REPORT

January 1, 1975 to December 31, 1975

BY: LEO VAN DUSEN
DIRECTOR OF PUBLIC WORKS

The San Bruno CATV system has now been in operation for four years. Construction of the cable system has been completed in Pacific Heights and Portola Highlands subdivisions. CATV is now available throughout 100% of the City. The total cable mileage is as follows:

Phase 1 - 2 - 3 Area	46.6 Mi.
Phase 4 Area	15.9 Mi.
Phase 5 Area	10.9 Mi.
	73.4 Mi.

With completion of the cable system, we can look forward to the expansion of the quality and type of service supplied by the system. Construction has already begun for a remote Head End Building on Sweeney Ridge, west of Portola Highlands subdivision. This remote Head End will allow us to pick up off the air additional television stations such as Channel 40 from Sacramento and provide better quality signal for the existing 4 Sacramento stations which we presently carry.

Pay television is presently in its beginning stage, throughout the United States, however, it offers a tremendous future potential. The capital cost for each of the various systems is large and must be thoroughly investigated before choosing the security system and firm to supply the programming.

The following television channels are presently carried by the system:

CATV Channel	Off Air Channel	Affiliate	City
Ch. 2	Ch. 2 - KTVU	Ind.	Oakland
Ch. 3	Ch. 3 - KCRA (Local Origination)	NBC	Sacramento
Ch. 4	Ch. 4 - KRON	NBC	San Francisco
Ch. 5	Ch. 5 - KPIX	CBS	San Francisco
Ch. 6	Ch. 14 - KCSM	PBS	San Mateo
Ch. 7	Ch. 7 - KGO	ABC	San Francisco
Ch. 8	Ch. 31 - KMUV	Ind. (F.L.)	Sacramento
Ch. 9	Ch. 9 - KQED	PBS	San Francisco
Ch. 10	Ch. 36 - KGSC	Ind.	San Jose
Ch. 11	Ch. 11 - KNTV	ABC	San Jose

Ch. 12	Ch. 20 - KEMO	Ind. (F.L.)	San Francisco
Ch. 13	Ch. 44 - KBHF	Ind.	San Francisco

Midband Channels

Ch. A	Ch. 3 - KCRA	NBC	Sacramento
Ch. B	Ch. 10 - KXTV	CBS	Sacramento
Ch. C	Ch. 38 - KVOF	Ind. (Rel.)	San Francisco
Ch. D	Ch. 13 - KOVR	ABC	Stockton
Ch. E	Ch. 60 - KDTV	Ind. (F.L.)	San Francisco

Future Channels

Ch. F	Ch. 26 - KTSF	Ind. (F.L.)	San Francisco
Ch. G	Ch. 32 - KQEC	PBS	San Francisco
Ch. H	Ch. 40 - KTXL	Ind.	Sacramento

SERVICE STATISTICS

Total Residential Connections	4,233
Total Commercial Connections	3,136
Total Secondary Connections	1,627
Midband Channel Converters	84

SERVICE PENETRATION

Based on the Water Department's water meter connections, the following are calculated percentages of service penetrations, based on the Federal Communications computation method of proportioning commercial and single family service connections.**

Location	Buildings	Service Connections	Penetration Percentage
Area 1 (Rollingwood, Monte Verde)	1,757	554	31.5%
Area 2 (Crestmoor)	1,638	1,142	69.7%
Area 3 (Freeway 280 to ECR)	2,850	1,497	52.5%
Area 4 (ECR to 7th Ave.)	3,169	1,429	45.1%
Area 5 (Portola Highlands, Pacific Heights)	1,394	271	19.4%
	10,808	4,893	45.3%

***FCC Method*

$$\frac{\text{3,136 Commercial Conn. @ \$1.00/Mo}}{\text{\$4.75/Mo Charge}} = \text{660 equiv. conns.}$$

DISCONNECTIONS

	1973	1974	1975	Totals	Percentage
Moving	442	626	700	1,768	63.9%
Dissatisfied	27	6	17	50	1.8%
Same reception	91	57	29	177	6.4%
Too much money	41	22	14	77	2.8%
Set problems	45	31	25	101	3.6%
Non Payment	101	101	93	295	10.6%
Not enough channels	8	2	0	10	0.4%
Miscellaneous	131	90	71	292	10.5%
	886	935	949	2,770	100.0%

Note: Some of the most common comments for Miscellaneous disconnections are temporary vacation, does not watch T.V., no T.V. set, died, getting divorced, Sutro Tower and other unknown reasons.

FINANCIAL OPERATIONS

Expenditures (January 1, 1975 to December 31, 1975	
Maintenance and Technical Salaries	$90,120
Administrative, Clerical and Sales Salaries	23,648
Materials and Supplies	61,450
Program Origination Services	1,025
Loan Interest	9,000
Pole Rental	5,924
	$191,167

REVENUES	$279,533
Less Expenditures	191,167
OPERATING PROFIT	$88,366
CAPITAL INVESTMENT	
Head End Facilities	
Land	$24,000
Buildings	27,922
Parking Area and Fencing	15,346
Electronic Equipment	41,150
Antennal Towers	19,175
Local Origination Equipment	14,700
T.V. Monitors	1,536
Total at Head End	$143,829

Equipment	
Test and Field Equipment	$ 2,456
Construction Equipment	6,478
Vehicles	22,336
Mobile Radios	3,563
Office Equipment	3,620
	$38,453

Distribution System	
Cable Plant	$476,444
Service Drops	131,565
Pole Clearance and Tree Trimming	74,079
Total Distribution System	$682,088
Total Capitalized Plant Cost	$864,370

FRANKFORT, KENTUCKY

Unlike San Bruno, whose City Council authorized construction of its cable system with city funds, Frankfort obtained its cable television system through its Municipal Electric and Water Plant Board, a semiautonomous organization within Frankfort's city government. Besides its obvious distinction as the first state capital (population, 22,000) to have its own public system, the cable operation–Community Service, Inc.–is managed by a unique nonprofit corporation establish to insulate the system from city politics.

Community Service has been extremely competent in extending cable service throughout the city: monthly rates are a very low $2.50 a month ($5.00 for installation). However, aside from a monthly audit performed by the Municipal Board and a five-year limitation on the service contract between Community Service and Plant Board, there is no direct citizen control over the activities of the Community Service directors–who, despite their managerial competence, were not elected or chosen through other democratic means.

A major controversy in Frankfort–seized upon by critics of publicly owned cable systems–has been the overwiring of a private operator by the Community Service system. As Ben Fowler, one of the original directors of Community Service, has explained, in a letter of May 15, 1974, the private operator had no authorization to operate in the city, though he continued to use municipal power poles by sufferance of the Plant Board. Apparently, citizen demand for access to Community Service simply compelled it to offer its service in those areas formerly served by the private operator. (No revenues are received by the city from the private operator.) Municipal power was not exercised to compel the private operator to cease operations at any time.

Equally controversial have been complaints (evidently originating with the private operator) alleging restricted programming on the municipal origination channel. As Fowler reported in a note accompanying the following letter, "Both systems operate a local channel, and each has been completely free to operate the local channel in any way it saw fit. There have been no restrictions of any kind on either system's programming."

The Frankfort system is strictly conventional in design and services, except for its excellent service and its unusual management structure. Ironically, the success of Community Service in providing a high-quality, conventional cable television system for Frankfort could discourage public debate, leading to possible future cable communications of a more innovative variety.

COMMUNITY SERVICE, INC.
Community Television Distribution System
325 ANN STREET
FRANKFORT, KY. 40601

June 30, 1972

Mr. Leslie T. Allen
City Manager
City Hall – P. O. Box 130
Bowling Green, Kentucky 42101

Dear Mr. Allen:

Mr. Allen Alsip, the acting City Manager of Frankfort, has referred your letter of February 4 to me. I was one of the original six directors of Community Service, Inc., which is the TV cable system owned by the city's electric company. I am still associated with the company and am presently serving as president.

In 1952 the Frankfort Electric & Water Plant Board, which operates the municipal electric and water systems, decided that cable TV was the only feasible method of television reception in the river valley area of Frankfort. Since it owned the poles and had access to those of the telephone company under a cooperative agreement, the Board determined that it would cause a TV cable system to be constructed and operated.

The Board asked six business and professional men in the City to form a non-stock, non-profit company to operate the system. This was done, and the Board initially constructed a system in a small area of the town. It entered into a contract with the newly formed company. Under the terms of the contract the company operates the system and manages the business independent of Plant Board control. However, the contract requires that the company must remain organized in such a way that no person has any ownership or right to profits derived from the operation, and any surplus funds can only be used for some local charitable or civic purpose.

The system has been expanded to cover all the city except an area in East Frankfort where private cable is in operation. The expansions have been financed by the company and no other funds have been advanced. Under the contract the legal title to the entire system remains in the municipal board.

The relatively small amount initially expended by the Plant Board was repaid in a few years.

The operation has been a success here in the sense that good TV reception has been provided to the valley area of the community and over the years we have developed an extremely low rate for the service. Up until October of 1967, our installation charge was $5.00 and our monthly service charge $2.00. In October we determined to completely rebuild the system and for that purpose we raised the monthly rate to $2.50.

The advantage of the arrangement made here, in my judgment, is that the operation although city owned is several steps removed from local politics. It also takes the burden of a rather difficult operation from the municipal power company which has sufficient problems of its own. The Municipal Board has sufficient supervisory control in that it requires an independent audit be submitted to it each year. The term of our current contract is five years, which means that in a relatively short period the entire set up can be reviewed by the Municipal Board in its consideration of granting a new contract.

The city itself has taken no action concerning the TV cable since it is in reality a project of its own Plant Board. The obvious advantage is to the citizens of Frankfort who receive the service for $2.50 per month when in most places the charge is $5.00 per month or higher. Also, the company is able to support the Chamber of Commerce and the United Fund very generously, as well as make sizeable contributions to various civic projects.

Most cities have issued franchises to TV cable companies. The Federal Communications Commission has, or is in the process, of placing a severe limitation on the amount of income a city may receive from a TV cable franchise. In my judgment the amount received by a city does not justify the issuance of a twenty year franchise which is a license to a cable company to receive an exorbitant fee from the people of the community for the service. TV cable rates are not regulated as are the other utilities. A company operating in Frankfort, where the use of the cable is almost universal, could recover its entire investment in two to three years if it charged $5.00 per month, which is the usual rate.

Unless the City of Bowling Green has a municipal electric system, I believe you would have considerable legal problems in trying to establish a system such as we have. I doubt if a city in Kentucky has the authority, under present laws, to expend money, or issue bonds, for a TV cable system. However, I assume you could issue a franchise which required an operation similar to this one. No private company would bid on such a franchise. It would require some local civic minded men to form a company and undertake the project. It is possible to obtain financing from some of the leading TV cable suppliers.

If I can be of further assistance call on me.

Yours very truly,

Ben B. Fowler

BBF:jht

PUBLIC CABLE TELEVISION AUTHORITY (NORTH ORANGE COUNTY, CALIFORNIA)

The most ambitious cable communications system in the nation, the result of more than five years of planning, is that being developed for the Public Cable Television Authority, a joint-powers agency comprising the cities of Westminster, Fountain Valley, Huntington Beach, and Newport Beach. Coordinating the affairs of the PCTA is John Bateman, independent communications consultant and president of John Bateman Associates. The PCTA system is planned as a regional telecommunications utility, incorporating newly constructed cable systems in Westminster, Fountain Valley, and Huntington Beach, and existing cable systems in Newport Beach.

According to the system design outlined in Bateman's presentation to the California Legislature Joint Committee on Telecommunications, which follows, a conventional cable television operation will share the physical plant with the Authority's own Community Service Communications System (CS2); the combined services will attract greater interest among the citizens of the PCTA cities and provide a mix of entertainment and information that has traditionally been missing from commercial cable television systems.

The CS2 concept is Bateman's innovation: in Orange County, where reception of television signals is acceptable—but where only one local station, a public broadcasting affiliate on the UHF band, programs for the community—the mixed system will allow the substantial development of local origination and public information services, which the PCTA cities have indicated is a priority need. Bateman, as part of the early planning for the system, canvassed the major educational and governmental institutions within the four PCTA cities in order to design a system "with the citizen in mind."

Bateman believes the CS2 component of the PCTA system will be exempt from FCC regulation: as an operationally independent cable system, carrying no broadcast signals, CS2 service standards will be specified by local authorities. Since CS2 is envisioned as an interactive utility, citizen groups will be able to participate in its programming and service development. Additionally, the availability of a regionwide two-way data transmission facility (via CS2) will strengthen the PCTA cities' competitive position in seeking to attract commercial enterprises—an especially important benefit in Orange County, the new center of information-based industry in Southern California.

At a meeting of the PCTA held on October 19, 1975, representatives of W. L. Dickinson & Co., management consultants retained by the PCTA, reported that they had devised an innovative plan for the financing of the system. According to the Dickinson consultants, the PCTA would create a nonprofit, municipal corporation to issue bonds in its own name, releasing the PCTA—an informal political entity—from direct financial responsibility for construction and operation of the cable system; this also would place a buffer between the political participants in the PCTA and the actual functioning of the system. (Of course, this separation would be largely ceremonial as long as the PCTA officials—members of the various city councils—were empowered to name directors on the board of the new corporation. The resulting accountability to the clients of the corporation would be infinitely preferable to the accountability to remote corporate moguls, who govern most commercial cable systems.)

In addition to issuing tax-exempt bonds, the corporation would contract with an operating company, possibly the Dickinson firm itself, to manage the cable system. The operating company would, in turn, locate investors willing to underwrite the revenue bonds financing the system; from revenues received, the operating company would periodically repay the debt to the investors. When the bonds were finally fully repaid, and with the termination of a 15-year contract between the corporation and the operating company, the corporation would be dissolved: ownership of the system would totally revert to the PCTA, which would authorize the operating company to manage the system or assert its responsibility in some other manner.

By January 1977, the PCTA had already accomplished necessary preliminary engineering studies (with $400,000 raised by Dickinson from local investors). A large bond offering is now before the Internal Revenue Service (IRS) for review and approval as a properly tax-exempt issuance. One likely consequence is the creation of not one but three municipal corporations, one for each of the cities of Westminster, Fountain Valley, and Huntington Beach. (Newport Beach, already partially cabled, is holding further construction in abeyance.) Upon receipt of IRS consent, the three corporations will issue municipal bonds enabling the cities to begin construction of an interconnected, jointly operated cable communications system; an initial offering of $7 to $8 million will be a concrete test of the validity of the PCTA concept.

Bateman's personal enthusiasm has propelled the PCTA toward accomplishment of its goals. Local schools and libraries have already begun to establish community production centers and program collections with which to fill the CS2 channels. The operation of an experimental two-way cable system in the schools of an adjacent city has no doubt contributed to a heightened awareness of the potentialities of cable communications for the citizens of the PCTA communities. While not engaging in an intensive public education program, Bateman has been careful to include the local press in all meetings of the PCTA—though it is becoming evident that greater citizen awareness cannot be put off indefinitely. A misleading headline in a regional daily paper, for instance, could have confused the citizens (whose taxes, after all, underwrite the revenue bonds) had not the article that followed explained the reason for issuing municipal bonds and outlined their benefits for the PCTA.

The Public Cable Television Authority is a bold and novel attempt to redefine the nature of the city and its responsibility to its citizens. It promises to make possible an urbanism that transcends traditional jurisdictional boundaries, as well as a technological infrastructure that will complement the PCTA cities' information-oriented economies. Its ultimate success, however, will depend on more than the virtuosity and competence of Bateman and his support staff in assessing the needs of the PCTA communities and providing solutions for those needs: the local citizens, the clients of the cable system, will vouch for the soundness of Bateman's and the PCTA's ideas when they begin to utilize—and, eventually, direct—the operating cable system.

PCTA/PUBLIC CABLE TELEVISION AUTHORITY
A Joint Powers Agency
8200 Westminster Avenue/Westminster, California 92603/(714) 893-4511

April 9, 1974

John R. Bateman
Consultant

Frank G. Fry, Jr.
Mayor Pro Tem Westminster
Chairman

Bernie P. Svalstad
Councilman Fountain Valley
Vice Chairman

Alvin M. Coen
Councilman Huntington Beach
Director

PRESENTATION BY: JOHN R. BATEMAN

TO: CALIFORNIA LEGISLATURE – JOINT COMMITTEE ON TELECOMMUNICATIONS
April 9, 1974, State Building,
Los Angeles, California

Chairman John Quimby; Vice Chairman, Fred Marler; Assembly Members,

Milam M. Dostal
Councilman Newport Beach
Director

Robert Cline and Walter Karabian; and Senate Members, Mervyn Dymally and Arlen Gregorio.

Opening Address

My name is John Bateman; I head up, as Consultant and Coordinator, the Public Cable Television Authority. This is a Joint Powers Agency, constituted under California law, comprising four West Orange County cities; namely, Fountain Valley, Huntington Beach, Newport Beach and Westminster.

I also have my own CATV consulting company, John Bateman Associates, which works exclusively for cities, counties, States, citizens and education groups. No manufacturers or cable operators are accepted as clients.

As Executive Director of the Public Cable Television Authority, I often speak on behalf of the PCTA Board of Directors. Today some of my comments must be divorced from such representation activity. In some areas, I will be speaking to you as an independent municipal cable consultant, specializing in city problems. Many of the views expressed will be my own, and in no way necessarily reflect the views of the PCTA Directors.

Wearing either of my hats, for the PCTA or John Bateman Associates, the objectives and goals for either group are almost identical: to develop sound, economic cable systems capable of providing a fair rate of return to the investor, thus insuring a continuing operation, simultaneously with the ability to provide much needed community programs and services to the communities, municipalities, educators, and the general public. Both the PCTA and John Bateman Associates place particular emphasis on the needs of the disadvantaged and the handicapped groups within the communities.

Your joint committee asked me to provide a brief description of the structure of the Public Cable Television Authority, its present status [in 1974] and future plans. You have also asked for suggestions as to the role the State might play as CATV develops within the State. Addressing myself logically to these questions, I will first cover the background of the Public Cable Television Authority.

History of the Public Cable Television Authority (PCTA)

In 1970, cities were faced with growing pressures from the cable industry to franchise cable operations. The cable operators were giving completely different stories, benefits and promises to the cities. A group of West Orange County cities banded together to explore their mutual problems. In October of 1971, they came up with a format to answer the problems and develop mutual benefits for the cities. On that date the cities of Fountain Valley, Huntington

Beach, Newport Beach, Westminster and Costa Mesa, the latter who at that time was a member, formed the first joint power CATV agency in the U.S.A., the Public Cable Television Authority. In March of the following year, 1972, I joined the PCTA as Consultant/Coordinator and filled the operational role as Executive Director.

Status of PCTA

What has the PCTA achieved since then?

a.) It developed an independent philosophy which could allow the cities to build and construct independent cable distribution systems, while at the same time share the advantages of a joint, overall system, which would transcend all city and community boundaries.

b.) It brought the seven (7) school districts representing 122 schools, 130,000 students together to work for common objectives and goals.

It further brought together our libraries for a common front effort.

c.) It designed a common overall engineering system, while maintaining each member group's geographic sovereignty.

d.) It developed legal ordinances and liaisoned them through each city's attorney.

e.) It published a 157 page request for proposal, which contained technical specifications and the first-ever detailed studio production specifications as a guide line for better quality origination.

f.) It developed a series of community programs across the board to help, assist, educate and serve groups of all ages, backgrounds, handicaps and needs.

g.) It investigated and devised ways to clear many of the FCC legal blocks to joint municipal, across the city boundaries, community benefits.

h.) It set the first standard and guide lines in the country, written with citizen benefits and needs in mind.

i.) It emphasized programs and services for the disadvantaged.

j.) It researched, objectively, the pros and cons of public ownership vs. franchise; not of the operating program arm of the system but of the actual signal carriage distribution part of the system.

k.) It developed the rationale to allow public ownership, if desired, and divorce political control from the operation. It developed the economic philosophy which would allow the *local businessmen*, the community and State to reap the financial rewards and benefits of such a system.

l.) It has developed the liaison and is coordinating various sections of the community which will allow time shared significant local community programming to develop, without the financial burden falling on the cable operator, as is the case in the current conventional CATV mode of today.

m.) It brought together joint ventures which will enable public safety factors, police, fire and paramedic departments to share and enjoy joint benefits for the overall enrichment and security of the associated PCTA member communities.

n.) All the above was created and balanced against the then existing realistic social-economic-demographic environments of the PCTA member cities. Such balances are constantly under review as the changing face of CATV finances and economics move constantly forward.

At the time the PCTA put the completed specifications out for bids, we received replies from some twenty cable operators who were interested. By the time we narrowed this to the few companies really *capable of delivering the specification requirements*, we were down to less than a handful. Having finalized on one company, we were then ready to proceed. At this point in time, the chosen cable operator's stock was suspended and we were left with no proposals. Starting from scratch again, since that time, we have developed four proposals: one from a banker, which would possibly allow us to build a public ownership system with revenue bonds, without city guarantees, and three other proposals for straight franchise. There are indications of other companies beginning to show a renewed interest. The close-off date for receiving proposals is next week, the 15th of April, 1974. Shortly thereafter the PCTA expects to make a decision on one of them. The PCTA reserves the option to reject all proposals if the need becomes apparent. [The PCTA did subsequently reject the proposals, preferring instead to create its own operating company with the help of the Dickinson firm. See the preceding introduction to this document.]

Background to Two New Approaches to Cable

In terms of what suggestions the State role might play in CATV development in California, I would like to interject two novel approaches for further detailed consideration. Time and space does not allow a full exposure of either approach.

Among the papers I have provided you with is a copy of the presentation I made to the Southern California Association of Governments (S.C.A.G.) on the 7th of February, at the LeBaron Hotel in Buena Park, where I have outlined the state of the industry and the problems the traditional CATV operating companies are having in obtaining capital for expansion.

After careful study of the impact of public ownership of the distribution section of the cable and the potential benefits it could have on local business communities, we are currently completing a matrix decision graph, which brings the pros and cons of such a public ownership out. It becomes fairly clear, when one studies the facts as opposed to the emotions often expressed by the cable industry, that all parties including the cable operators have a great deal to benefit from public ownership of cable distribution systems.

It is a mandatory need that the municipal aspects of such an ownership are confined to the actual overall plant distribution operation. All programming and channel operations are franchised out to different community groups and interests, under a normal free enterprise competitive system.

The implications and benefits that such a marriage of public and free enterprise systems can have on the community is immense. If CATV systems are franchised to owners outside of the State, then a significant part of the revenue generated from the communities goes out of the State to the head office of the cable operations, and eventually into the coffers of other State operations. Advertising tends to accent certain national connotations rather than assist the development of the local business merchants through the ability to provide a cheap advertising rate for local origination.

We believe that local distribution ownership, with the generated revenue dollar recirculating within the business communities, could well circulate eight to a dozen times before finally leaving the community.

On a public ownership system, which looks purely at the ownership of the distribution system, there is a distinct advantage to the local business community which would certainly benefit most. If it is true, from this hypothetical position, that we can really benefit our communities by public ownership of just the distribution system, while leaving the programming, management and other aspects of the system to the free enterprise and local communities, what are the problems involved in developing such a system?

State Support for Public Ownership of Distribution System

Currently, few cities in California have the credit capability of underwriting a revenue CATV bond issue, and whether such a question is even legal could be raised. At the present time, a CATV system developed on revenue bonds would not have confidence of the public, unless underwritten and guaranteed; i.e., by a responsible fiscally sound body.

One possible approach would be to examine the possibilities of the State, on an experimental basis, underwriting the revenue bonds after first building in adequate safeguards for an experimental system. This would allow, under rigid controlled conditions, a chosen city or group of cities to obtain revenue bond financing to build a CATV distribution system, and lease out the operational elements to industry. The debt service would be provided from the system subscribers and also revenue generated from leased channels and leased services.

In this approach, the shortcomings in debt service, even if looking at an overall system failure, would be marginal and recoverable over a period of years. In the very worst adverse conditions, during the first five or six years of the operation, there might be a deficiency of gross revenue to cover the debt. If the State could guarantee the debt service to the public on even one well-balanced pilot system, it would be possible to build a general communication medium designed for general communications and services for the use of the communities, the public safety elements, the educational elements and local government operations. This is a cognizance of what I have said without looking further into the mass of details and complications involved; but there is substance for what I am proposing, and I would strongly urge you to examine the possibilities and details necessary to instigate such an experiment.

The success of such a system would fill the void of cable operators who are unable to generate the capital investment necessary to do the job needed of cable operators in today's environment. It would allow local business people to generate local business, it would allow the generated revenues to stay within the communities, or at least mainly within the State; and it would generate more job openings within the State, while at the same time develop needed community services.

Providing a Low Price Community Cable System

I should like to make a second suggestion. Rich Muller, one of my principal consultants with John Bateman Associates, and a recognized CATV marketing research authority, has indicated to me time after time, the systems that are being proposed, and currently are being built in this country today, are too complicated. More and more they are becoming highly expensive and difficult, if not impossible to capitalize. The Federal regulations have reduced the length of the franchise contract to a point where amortization of capital investment is almost impossible; and for a really modern two-way system, it is totally impossible.

There is an answer. Referring back to my S.C.A.G. paper, the level of services to a community can be structured at a three-fold level, and if I may state the problem in a logical manner, it is thus:

The primary traditional approach to establish a CATV system is to provide a service to the public of ten to fifteen or more broadcast channels. Its second mandatory duty is to accommodate local origination programs. Its third duty should be to generate public services.

The cable operator then charges from $5.00 to $7.00 for the TV reception service, or whatever he thinks the market will bear for the minimal services. Obviously, these prices are extremely sensitive from market to market with regards to the aggregate total billing the communities are capable of paying for such a system. In the type of major cities we are looking at today, this is extremely important because our educational and community groups have great needs of cable services, which are capable of providing service to the community groups. These cover a very large range of service such as: paramedics, fire, police, security, meter reading, information, community programs and other services which could improve the quality of our daily lives.

In many of our open markets today, we do not require cable systems to bring in TV stations. With good antennas in the Los Angeles and general area, we can bring in seventeen or eighteen stations today. Cable is really marginal for anyone who wants to take the trouble of erecting good antennas and pointing them in the right direction.

The solution is simple. Do not build the traditional CATV system in the marginal market. Instead, establish a community service communications system (CS2), the fundamentals of which would be a number of channels going down stream, not carrying broadcast signals but community services and local origination, the education, the public access, and the municipality and community information programming. Channels could be provided for lease for two-way considerations covering a whole range of the services required in the emerging markets of our domestic and business communities. Such a system would do away with the highly sophisticated, complex, multi-channel equipment necessary in a modern two-way, fully operational cable system. It would do away with the complicated cable heads and links; it would simplify the maintenance problems and costs; it would reduce drastically the overall cash requirements necessary to invest in starting up a system. It would enable the subscriber, with a simple A/B switch box, to switch his normal UHF/VHF channels to receive his regular off-air pictures, or to place him on the cable and substitute those ambient signals. There are technical problems to this, but these are entirely solvable.

The cost advantage becomes enormous. By reducing the capital investment and maintenance costs, and combined with a high level of community involvement inherent in the type of system I am talking about, we will be generating services and programs unavailable elsewhere, at a price and quality acceptable to many of the potential subscribers.

Such a system could economically provide top-rate needed community services at something like $3.00 per month.

To be able to support this contention, we must be able to generate the software programming from within our own community sources.

Gentlemen, I don't know what is available in many of the other areas of the State, but I know within my cities, my schools alone have something like $600,000 worth of production equipment, and much of this is sitting idle or incorrectly used. This equipment could be utilized through the schools for their *own benefit* to carry educational programming into the homes. Through local origination studios and through use of our libraries, we can generate the software programming required, and with the help of such organizations as the California Association of the Blind, the clinics for the impaired hearing, groups dedicated to the mentally retarded, the mentally trainable, the physically handicapped, and through such groups as the senior citizens, a whole range of software programs can be organized, brought together and established for the good of the community.

All of this can be done still utilizing the free enterprise system of allowing the community business groups to operate the commercial activities, less the off-air reception aspects, without the necessity of meeting the overall cost of unprofitable, for them, community programming.

The legal questions of whether the FCC's 1934 Communications Act would even enter into the picture is a wide one, and there are many attorneys capable of arguing this; but it would resolve in developing cable systems which could benefit our community today, to look after them, to help them, to settle and solve many of the problems besetting our communities today; and it could be done at a cost which could be available for nearly all to subscribe to, as opposed to the very complex systems which very few can afford or even need in our top one-hundred markets today.

The benefits to the State and community do not stop there. With a high potential of subscribers, on a low-cost system, within a given community, we would very rapidly be looking at an entirely wired nation, which would have the capability of providing a large range of services to the community.

How perhaps could we achieve this? We could achieve this in exactly the same way as I proposed earlier: i.e., by the State setting aside, for an experimental expenditure, a sum of money which would allow the State to guarantee, if all else fails, the debt service of a revenue bond for such an experimental cable system in a given, chosen city or group of cities. Once such a demonstration proved financially possible, it would open up the whole communication ability nationwide.

To conclude, Gentlemen, I would like to stress—in California we have high seats of learning and education, we have innovative local government, we have a progressive Assembly and Senate, we have the opportunity, in-house capability, of piloting such a system which would blow the lid off CATV, as we know it today, and allow an ultra-modern model communications development.

It would benefit businessmen, educators and government alike. Such a system would possibly limit the capital gains of the entrepreneur, but would open even greater markets for cable operators, businessmen, job opportunities, and provide needed community services.

Chairman Quimby, Assembly members, Senate members, your letter to me asked for positive suggestions on the role the State might play as CATV develops. Please keep an open mind and consider what I have proposed today.

POSTSCRIPT TO:
PRESENTATION BY: JOHN R. BATEMAN
TO: CALIFORNIA LEGISLATURE – JOINT COMMITTEE ON TELECOMMUNICATIONS
April 9, 1974, State Building, Los Angeles, California

After the close of the LEGISLATIVE HEARINGS, a number of people sought clarifications on the simplified CS2 proposal.

The CS2 proposal does not seek to reduce or limit in any way the important role of the Public Broadcast Station. On the contrary, the proposal seeks to utilize to an even greater effect the interchange of multiple diverse programs now being exchanged between local and State colleges and the local Public Broadcast Stations.

The intent is to exchange programs and reoriginate the college and PBS programs on special full-time dedicated community cable (CS2) channels. These programs can be run and rerun around the clock, which will allow far greater exposure to the student and the general public at hours that are convenient to the individual viewer in the community.

Students studying certain courses will be able to view in the evening or during the night those programs that have been missed during the day. The week's programs can be reshown on Saturdays and Sundays, allowing serious students to catch up on missed content during the week. Rerun schedules will be obtainable by telephone from the CS2 offices.

Such flexible schedules must provide optimum coverage of all programs and further should allow the greatest possible access to students studying for college credit.

Through such methods, by use of non-off air cable services at low cost to the community, available to all, the true "University of the Air" concept can reach out into the homes and offer true equal opportunity for all.

The same method of approach can reach out and help all handicapped and disadvantaged special interest groups. It can further provide help and services to the crowded cities, high-ambient-TV signal areas and the rural areas, the latter requiring greater state or national funding help. All of these special areas, in many parts of the country, currently have little chance of ever being able to achieve programs and services under the existing financial patterns used by cable operators today. Yet even those operators in most cases are willing to help, if alternate methods of finance can be found so that they alone do not have to carry the burden.

BIBLIOGRAPHY

Achtenberg, Ben. "A Word to the Wise: Study the Franchise." *Urban Review* 6, no. 2 (November-December 1972): 8–12.

Ackoff, Russell L. "Toward Quantitative Evaluation of Urban Services." In *Public Expenditures Decisions in the Urban Community*, ed. Howard Schaller, pp. 91–117. Washington, D.C.: Resources for the Future, 1963.

American Civil Liberties Union. *Policy #20, Cable Television*. Washington, D.C.: ACLU, July 22, 1968.

"The Anaheim-Arlington Connection." *Broadcasting* (August 5, 1974): 48–49.

"The Ant's-Eye's View of Architecture." *Human Behavior* 3, no. 8 (August 1974): 48.

Ash, Maurice. *Regions of Tomorrow: Towards the Open City*. London: Evelyn, Adams & Mackay, 1969.

Baer, Walter S. *Cable Television in the United States–Revolution or Evolution?* Santa Monica, Calif.: Rand Corporation, 1974.

Bagdikian, Ben. H. *The Information Machines*. New York: Harper & Row, 1971.

Baran, Paul. *Potential Demand for Two-Way Information Services to the Home, 1970–1990.* Menlo Park, Calif.: Institute for the Future, December 1971.

Barnes, Peter, and Shearer, Derek. "Beyond Antitrust." *The New Republic* 172, nos. 1–2 (July 6 and 13, 1974): 16–19.

Barnouw, Erik. *A History of Broadcasting in the United States*. Vols. 1–3. New York: Oxford University Press, 1966–70.

Berson, James H. "The Human Values of City Planning." *Journal of Environmental Systems* 1, no. 3 (September 1974): 283–87.

Bish, Robert L., and Warren, Robert. "Scale and Monopoly Problems in Urban Government Services." *Urban Affairs Quarterly* 8, no. 1 (September 1972): 97–122.

Botein, Michael. "Access to Cable Television." *Cornell Law Review* 57, no. 3 (February 1972): 419–59.

Includes works not cited in the text but relevant to the topic.

Branscomb, Anne W. "The Cable Fable: Will It Come True?" *Journal of Communication* 25, no. 1 (Winter 1975): 44-56.

———. "Killing the Goose that Lays the Golden Egg," *Journal of Communication* 25, no. 3 (Summer 1975): 221-22.

Briscoe, Wally. "Cable: Big Brother's Victim–Not His Tool." *TV Communications* (reprint, December 1972).

"In Cable Franchising, the Cards Are Being Played Closer to the Vest." *Broadcasting* (February 4, 1974): 48-51.

"Cablecasting: A Myth or Reality?" *Rutgers Law Review* 26, no. 804 (Summer 1973): 804-37.

Cable Television Information Center. *Local Government Uses of Cable Television.* Washington, D.C.: The Urban Institute, 1974.

"Cable-TV Dangles New Lures." *Business Week* (December 1, 1973): 63-64.

California Community Television Association. *For California Legislators . . . What Cable Television Can Do for You.* Castro Valley, Calif.: California Community Television Assn., n.d.

California Council of Intergovernmental Relations. *Local Government Planning in California.* N.p., 1972.

Calvert, Gordon L., ed. *Fundamentals of Municipal Bonds.* New York: Securities Industry Association, November, 1973.

Campbell, Alan K., and Sacks, Seymour. *Metropolitan America.* New York: The Free Press, 1967.

Catanese, Anthony James. *Planners and Local Politics: Impossible Dreams.* Beverly Hills: Sage, 1974.

"CATV of, by and for the People." *Broadcasting* (March 11, 1974): 79.

Chermayeff, Serge, and Tzonis, Alexander. *Shape of Community.* Baltimore: Penguin, 1971.

Clay, Grady. *Close-Up: How to Read the American City.* New York: Praeger, 1973.

Community Video Center. *A Proposal for an Urban Video Network for the Benefit of the Washington, D.C. Community.* Washington, D.C.: Community Video Center, 1971.

"County-City Cable Plan." *Broadcasting* (February 25, 1974): 54.

"Court Strikes Down FCC Pre-emption of Regulation of Leased, Two-Way Cable Channels." *Broadcasting* (February 16, 1976): 54.

Dahl, Robert A. "Governing the Giant Corporation." In *Corporate Power in America*, ed. Ralph Nader and Mark Green, pp. 10-24. New York: Grossman, 1973.

Delos Four. "Need for More Balance in the Flow of Communications." In *Human Identity in the Urban Environment*, ed. Gwen Bell and Jacqueline Tyrwhitt, pp. 240–44. Harmondsworth, U.K.: Penguin, 1972.

Detroit, City of, Cable T.V. Study Committee. *Cable Television in Detroit*. Detroit: City of Detroit, 1972.

Dordick, Herbert S., et al. *A Study of Local Government Uses of Cable Television and Integrated Communication Systems*. Los Angeles: Center for Communication Policy Research, University of Southern California, 1975.

Dordick, Herbert S., and Pilnick, Carl. *Integrated Urban Telecommunications for the Delivery of Municipal and Social Services in Los Angeles–Design for System Interconnection*. Los Angeles: Center for Communications Policy Research, University of Southern California, 1975.

Dowlin, Kenneth E. *The NCPL Scene–1980 Script for Video Tape*. Casper, Wyo.: Natrona County Public Library, October 3, 1974.

———. *A Summary Report of the Activities of the Natrona County Public Library in Its Use of CATV to Provide Library Services*. Casper, Wyo.: Natrona County Public Library, December 13, 1974.

Elgin, Duane, et al. *City Size and the Quality of Life*. Washington, D.C.: National Science Foundation, 1974.

Enzensberger, Hans Magnus. "Constituents of a Theory of the Media." In *Sociology of Mass Communication*, ed. D. McQuail, pp. 99–116. Baltimore: Penguin, 1972.

Fainstein, Norman I., and Fainstein, Susan S. *Urban Political Movements: The Search for Power and Minority Groups in American Cities*. Englewood Cliffs, N.J.: Prentice-Hall, 1974.

"FCC Says It's Up to Congress to Define Three Tiers of Cable Regulation." *Broadcasting* (August 18, 1975): 27.

Finland, Government Committee on Communication Policy. *Investigations and Recommendations Concerning Developments in Communication Technology (Abridged)*. Helsinki: Finnish Council of State, 1974.

"Fla. Cable Fights with County over Proposal for Competing Translators." *Broadcasting* (March 18, 1975): 91.

Foster, David. "Aeropagitica Revisited." Speech before the Federal Communications Bar Association, March 15, 1973, Washington, D.C. Mimeographed.

Foundation 70. "A Cable Is a Very Big Wire." *Yale Review of Law and Social Action* 2, no. 3 (Spring 1972): 199-207.

Fowler, Floyd J., Jr. *Citizen Attitudes Toward Local Government, Services, and Taxes.* Cambridge, Mass.: Ballinger, 1974.

Friendly, Fred. "Remarks Prepared for Delivery." Speech delivered at PubliCable Seminar, May 5, 1972, Washington, D.C. Mimeographed.

Gabel, Richard. "The Early Competitive Era in Telephone Communication, 1893–1920." *Law & Contemporary Problems* 34 (1969): 340–59.

Gans, Herbert J. Letter in "The Planning Debate." *Social Policy* 6, no. 7 (May-June 1975): 25–27.

Gerbner, George. "Preface." In *Communications Technology and Social Policy*, ed. George Gerbner, Larry P. Gross, and William H. Melody, p. x. New York: John Wiley & Sons, 1973.

Glavitsch, Hans. "Computer Control of Electric-Power Systems." *Scientific American* 231, no. 12 (January 1975): 34–44.

Green, Mark, et al. *The Closed Enterprise System: The Report on Anti-Trust Enforcement.* New York: Grossman, 1972.

Gross, Bertram M. "Introduction [to Part IV]." In *Communications Technology and Social Policy*, ed. George Gerbner, Larry P. Gross, and William H. Melody, pp. 289–92. New York: John Wiley & Sons, 1973.

Hanson, Royce. "Opportunities for Innovative Local Government." In *The Contemporary New Communities Movement in the United States*, ed. G. Golany and D. Walden. Urbana: University of Illinois Press, 1974.

Head, Sydney. *Broadcasting in America.* New York: Houghton Mifflin Company, 1972.

Hénaut, Dorothy Todd. "A Few Notes on Regional Projects." *Access, Challenge for Change/Société Nouvelle*, no. 10 (Autumn 1972): 11.

Henderson, Hazel. "Information and the New Movements for Citizen Participation." *The Annals of the American Academy of Political and Social Science* 412 (March 1974): 34–43.

Hernandez, William. *An Exploration into the Issues of an Emerging National Policy for Cable Television.* Master's Thesis, University of California-Los Angeles, 1973.

Heyman, Mark. "What People Do in the City." *Planning* 1, no. 7 (August 1975): 26–27.

Himsworth, W. E. *Cable Television–An Industry Analysis.* New York: Salomon Brothers, February 29, 1972.

Hooper, Richard. "Yet More Spaghetti." *Urban Review* 6, no. 7 (November-December, 1972): 4–7.

Huddle, Franklin P. "The Social Function of Technology Assessment." In *Technology Assessment*, ed. Raphael G. Kasper, pp. 149–67. New York: Praeger, 1972.

Jacobson, Robert E. "'The Cable Fable': Lost in Valhalla," *Journal of Communication* 25, no. 3 (Summer 1975): 219–21.

Jencks, Charles, and Silver, Nathan. *Adhocism*. Garden City, N.Y.: Doubleday, 1972.

Kalba, Kas. "The Cable Fable." *Yale Review of Law and Social Action* 2, no. 3 (Spring 1972): 195–98.

———. "Telecommunications for Future Human Settlements: A Planning Framework for Minnesota Experimental City." *Ekistics* 35, no. 211 (June 1973): 329–36.

———. "Urban Telecommunications: A New Planning Context." *Socio-Economic Planning Science* 8 (February 1974): 34–45.

Kato, Hidetoshi. "From Pantheon to Presley: Changes in Urban Symbolism." In *Communication and the City: The Changing Environment*, ed. East-West Communication Institute, pp. 3–16. Paper no. 7. Honolulu: East-West Communication Institute, November 1973.

Katzman, Natan. "The Impact of Communication Technology." *Ekistics* 225 (August 1974): 125–30.

Kloman, Erasmus H. "Public Participation in Technology Assessment." *Public Administration Review* 34, no. 1 (January-February 1974): 52–61.

Kuenzlen, Martin. *Playing Urban Games*. Boston: i press, 1972.

LaRue, Steve. "The Small Cities' Threat to the Giant Utilities." *California Journal* 6, no. 9 (September 1975): 318–20.

Ledbetter, Ted. "Broadband Communications in an Urban Setting." In *The Wired City of Los Angeles*, ed. Robert Warren, pp. 3–12. Los Angeles: Center for Urban Affairs, University of Southern California, 1972.

LeDuc, Don R. *Cable Television and the FCC*. Philadelphia: Temple University Press, 1973.

———. "Cable-TV Franchising and the Local Bar Association." *Wisconsin Bar Bulletin* 47, no. 6 (December 1974): 38–43.

Lemieux, Barry. "Ownership Patterns and Alternatives." In *The Wired City of Los Angeles*, ed. Robert Warren, pp. 65–72. Los Angeles: Center for Urban Affairs, University of Southern California, 1972.

Lenz-Romeiss, Felizitas. *The City–New Town or Home Town?* New York: Praeger, 1973.

Leone, Richard C., and Powell, Robert S., Jr. "CATV Franchising in New Jersey." *Yale Review of Law and Social Action* 2, no. 3 (Spring 1972): 252–60.

Litterman, Bob. "Who Owns the Media?" *Pacific Research and World Empire Telegram* 4, no. 2 (January-February 1973): 9–16.

Little, John C., Chander A. Stevens, and Peter Tropp. "Puerto Rico's Citizen Feedback System." In *Analysis of Public Systems*, ed. Alvin W. Drake, Ralph L. Kinney, and Philip M. Morse. Cambridge, Mass.: MIT Press, 1972.

Long, Norton E. *The Unwalled City*. New York: Basic Books, 1972.

Los Angeles, Study Team on Franchise Districts. Los Angeles: Los Angeles City Department of Public Utilities and Transportation, February 17, 1976.

Los Angeles, Study Team on Service Needs. *Interim Report*. Los Angeles: Los Angeles City Department of Public Utilities and Transportation, January 15, 1976.

Lukács, Georg. *History and Class Consciousness*. Cambridge, Mass.: MIT Press, 1971.

Lyall, Katherine C. "Municipal-Service Applications of Cable: Some Caveats." *Bulletin of the American Society of Information Scientists* 2, no. 1 (June-July 1974): 28–29.

Lynch, Kevin. *The Image of the City*. Cambridge, Mass.: MIT Press, 1960.

McHale, John. "The Changing Information Environment: A Selective Topography." *Ekistics* 35, no. 211 (June 1973): 321-28.

McKean, Roland N. *Public Spending*. New York: McGraw-Hill, 1968.

Maitland, John C., Jr. "Ethics and Cable Television." *Audio-visual Instruction* (December 1973): 33.

Mandlebaum, Seymour J. *Community and Communications*. New York: W. W. Norton, 1972.

"Market Briefs: The Municipal Bond Insurance Assn." *Business Week* (February 16, 1976): 72.

Marris, Robin. "Businesses, Economics and Society." In *Social Innovation in the City*, ed. Richard S. Rosenbloom and Robin Marris, pp. 19–49. Cambridge, Mass.: Harvard University Press, 1969.

Meier, Richard L. *Communications Stresses in the Future Urban Environment*. Berkeley: Institute of Urban and Regional Development, University of California, 1968.

———. *A Communications Theory of Urban Growth*. Cambridge, Mass.: MIT Press, 1970.

Melody, William H. "The Role of Advocacy in Public Policy Planning." In *Communications Technology and Social Policy*, ed. George Gerbner, Larry P. Gross, and William H. Melody, pp. 165-81. New York: John Wiley & Sons, 1973.

MITRE Corporation. *Symposium on Urban Cable Television*. Vols. 1–4. McLean, Va.: MITRE Corporation, 1973.

Molenda, Michael H. "CATV and Access to Knowledge." *Yale Review of Law and Social Action* 2, no. 3 (Spring 1972): 243-51.

Moller, Clifford B. *Architectural Environment and Our Mental Health*. New York: Horizon Press, 1968.

Morris, David, and Hess, Karl. *Neighborhood Power*. Boston: Beacon, 1975.

Moore, William D. "Influences of State and Local Law." In *Cable Report for Jacksonville, Florida*, ed. Cable Television Information Center. Jacksonville, Fla.: Cable Television Information Center, addendum.

"More Local Programs with Better Quality Going Over the Cable." *Broadcasting* (September 9, 1974): 62.

Mumford, Lewis. *The Urban Prospect*. New York: Harcourt, Brace & World, 1968.

National Academy of Engineering, Committee on Telecommunications. *Telecommunications for Enhanced Metropolitan Function and Form*. Washington, D.C.: National Academy of Engineering, 1969.

Network Project, The. *Cable Television*. Notebook no. 5. New York: The Network Project, Columbia University, 1973.

———. *Cable Television: End of a Dream*. Notebook no. 8. New York: The Network Project, Columbia University, 1974.

"No Raise, No Service." *Broadcasting* (October 14, 1974): 42.

"In Oakland, a Cable-TV System Fails to Live Up to Promises, Irking City and Minority Groups," *Wall Street Journal*, September 25, 1974, p. 30.

Oppenheim, Jerrold N. "Cable TV and the Public Interest." *The Progressive* 38, no. 3 (March 1974): 46–50.

———. "The Unfulfilled Promise of Cable TV." *The Progressive* 38, no. 2 (February 1974): 49–52.

Othmer, David. *The Wired Island*. New York: Fund for the City of New York, 1973.

Parker, Edwin B. "Networks for an Information Society." *Bulletin of the American Society for Information Science* 2, no. 1 (June-July 1975): 12–14.

———. *On-Line Polling and Voting*. Mimeographed. Stanford, Calif.: Stanford University, 1971.

———. "Technology Assessment or Institutional Change?" In *Communications Technology and Social Policy*, ed. George Gerbner, Larry P. Gross, and William H. Melody, pp. 533–45. New York: John Wiley & Sons, 1973.

"Physician, Heal Thyself." *Forbes* (December 1, 1973): 62–65.

"Poverty Is Major Issue in Comments on '77 Deadline." *Broadcasting* (March 24, 1975): 38–39.

President's Cabinet Committee on Cable Communications. *Cable Report to the President, January 14, 1974*. Reproduced in *Television Digest* (January 21, 1974), special insert.

Press, Laurence I. "Arguments for a Moratorium on the Construction of a Community Information Utility," *Communications of the ACM* 17, no. 12 (December 1974): 672–79.

"Puerto Rico Disconnects ITT." *Business Week* (February 16, 1974): 52.

"Queens System Hunts Shortcuts to Quick Start." *Broadcasting* (March 24, 1975): 40.

Rabinowitz, Alan. *Municipal Bond Finance and Administration*. New York: Wiley-Interscience, 1969.

Raskin, Marcus G. "The Dream Colony." In *Television Today*, edited by Richard L. Stavins, pp. 5–28. Washington, D.C.: Communication Service Corporation, 1969.

Renfuss, John. *Public Administration as Political Process*. New York: Charles Scribner's Sons, 1973.

Rivers, William L., and Nyhan, Michael J., eds. *Aspen Notebook on Government and the Media*. New York: Praeger, 1973.

Rivkin, Steven R. *Cable Television: A Guide to Federal Regulations*. No. R-1138-NSF. Santa Monica, Calif.: Rand Corporation, 1973.

Rockwell, Arthur E. *Quarterly Survey of the Cable Television Industry*. San Francisco: Sutro & Company, January 30, 1974 and May 1974.

Roman, Daniel D. "Technology Assessment: Perspective from the Managerial Position." *Public Administration Review* 33, no. 5 (September-October 1973): 393–99.

Ross, Leonard. "Minibonds for Utilities: An Idea Whose Time Has Come?" Los Angeles *Times*, June 2, 1975, pt. II, p. 7.

Roth, Edward J. "Should City Hall Own the Cable TV System?" *Nation's Cities* 11, no. 6 (June 1973): 41–43.

Rusan, Francille, "What You See Is What To Get: Cable Television and Community Control." *Yale Review of Law and Social Action* 2, no. 3 (Spring 1972): 275–81.

Sackman, Harold. "Computers and Human Problems." In *Computers and the Problems of Society*, ed. Harold Sackman and H. Borko, pp. 1–30, Montvale, N.J.: American Federation of Information Processing Societies (AFIPS) Press, 1972.

———. *Mass Information Utilities and Social Excellence*. Princeton, N.J.: Auerbach, 1971.

Saldich, Anne Rawley. "Facts and Fables About Cable TV." *Television Quarterly* 11, no. 4 (Summer 1974): 49–53.

San Gabriel Valley Public Cable Council, Pasadena Urban Coalition, and Beverly Hills Bar Association Law Foundation. *Municipal Decision-Making Factors Relative to Cable Television Ownership*. Pasadena, Calif.: Pasadena Urban Coalition, 1973.

Sarason, Seymour B. "Community Psychology, Networks, and Mr. Everyman." *American Psychologist* 31, no. 5 (May 1976): 321–28.

Scherer, Jacqueline. *Contemporary Community*. London: Tavistock, 1972.

Schiller, Herbert I. *The Mind Managers*. Boston: Beacon, 1973.

———. "Waiting for Orders—Some Current Trends in Mass Communications Research in the United States." *Gazette* 20, no. 1 (1974): 11–21.

Schwartz, Louis, and Woods, Robert A. "A Marriage Proposal: Cable Television and Local Public Power." *Public Power* (January-February 1972).

Scott, Mel. *American City Planning Since 1890*. Berkeley: University of California Press, 1969.

Scott, Stanley. "Lessons from the History of Broadcasting." *Science* 178 (December 22, 1972): 1263–65.

Seiden, Martin H. *Cable Television U.S.A.* New York: Praeger, 1972.

"Senate Study Sees Big Bank Presence in Broadcast Firms." *Broadcasting* (January 14, 1974): 41–43.

Singer, Benjamin D. *Feedback and Society*. Lexington, Mass.: D. C. Heath & Co., 1973.

Smith, Ralph Lee. "CATV: FCC Rules and the Public Interest." In *Communications Technology and Social Policy*, ed. George Gerbner, Larry P. Gross, and William H. Melody, pp. 121–31. New York: John Wiley & Sons, 1973.

———. "Ownership Policy and the Cable Industry." *Yale Review of Law and Social Policy* 2, no. 3 (Spring 1972): 263–74.

Stedman, Murray S., Jr. *Urban Politics*. Cambridge, Mass.: Winthrop Publishers, 1972.

Steiner, Robert L. *Visions of Cablevision*. Cincinnati: Stephen H. Wilder Foundation, 1972.

Stern, Alfred R. Speech before the New England Cable Television Association, April 18, 1973, Hyannis, Mass. Mimeographed.

Stern, Joseph L. "Business Data on CATV Possible with Wideband Analog Cable Systems." *Communication News* 12, no. 1 (January 1975): 30–31.

Stewart, Walter. "Why Cities Don't Work." *MacLean's* 87, no. 1 (January 1974): 19.

Suttles, Gerald D. *The Social Construction of Community*. Chicago: University of Chicago Press, 1972.

Sweden, Commission for Radio and Television in Education. *Kabelvision Kuruna*. Government official reports SOU 1975:28. Stockholm: TRU-Committee, 1975.

Szanton, Peter L. "Systems Problems in the City." In *Rand 25th Anniversary Volume*, edited by Rand Corporation. Santa Monica, Calif.: Rand Corporation, 1974.

Tate, Charles. *Cable Television in the Cities*. Washington, D.C.: The Urban Institute, 1971.

Telecommunications Management Corporation. *Cable Communications for the City of Portland and Multnomah County*. Los Angeles: Telecommunications Management Corporation, 1973.

"Toward Community Ownership of Cable Television." *Yale Law Journal* 83, no. 8 (July 1974): 1708–29.

"Turning on Telepsychiatry." *Human Behavior* 3, no. 5 (May 1974): 30.

Tzonis, Alexander. *Towards a Non-Oppressive Environment*. Boston: i press, 1972.

Urban Institute, The. *The Struggle to Bring Technology to the Cities*. Washington, D.C.: The Urban Institute, 1971.

U.S. Conference of Mayors. *Mayors' Action Report on Cable Television*. Washington, D.C.: U.S. Conference of Mayors, 1973.

von Hertzen, Heikki, and Spreiregen, Paul D. *Building a New Town*. Cambridge: MIT Press, 1973.

Wallace, Edwin. "Cities Can Supply Cable Television Systems." *The American City* 87, no. 9 (September 1972): 95.

Washnis, George J. *Municipal Decentralization and Neighborhood Resources*. New York: Praeger, 1972.

"Warner Backs Away." *Cable Information Newsletter* 3, no 3 (March 1974): 2–3.

Webbink, Douglas W. "Should Cable TV Be Regulated as a Public Utility?" *Public Utilities Fortnightly* 89, no. 13 (June 22, 1972), pp. 35–38.

"White House Moving Ahead on De-regulation." *Broadcasting* (October 13, 1975): 26.

Williams, Donald O., and Kaitz, Spencer R. *Cities Should Not Own CATV Systems Because . . .* National Cable Television Association.

Yin, Robert K., B. L. Kinney, and K. B. Possner. *Neighborhood Communications Centers: Planning Information and Referral Services in the Urban Library*. No. R-1546-MF. Santa Monica, Calif.: Rand Corporation, 1974.

ABOUT THE AUTHOR

ROBERT E. JACOBSON, a doctoral student in the School of Architecture and Urban Planning at the University of California, Los Angeles, is a consultant with cities and community groups in Southern California on communication policy-related matters.

Before entering the doctoral program at UCLA, Mr. Jacobson was a doctoral student and research associate at the Annenberg School of Communications at the Annenberg School of Communications at the University of Southern California. He was instrumental in the initiation of a "cable master planning process" in the City of Los Angeles in 1975-76 and participated in it as co-chairman of the Study Team on Ownership of the Los Angeles Citizens Task Force on Cable Communications. Prior to his work on the Citizens Task Force, Mr. Jacobson conducted seminars on cable planning for the City of Santa Monica and the Southern California Association of Governments; additionally, he was a founding member of the Los Angeles Public Access Project, a non-profit community production center.

Mr. Jacobson completed his B.A. and M.A. degrees at UCLA; his thesis for the latter in Critical Studies/Television, evolved into the present study. Mr. Jacobson has also authored several articles and short pieces appearing, or scheduled to appear, in the *Journal of Communication, Instant Research on Peace and Violence*, and local periodicals.

RELATED TITLES

Published by
Praeger Special Studies

*THE MASS MEDIA: Aspen Guide to Communications Industry Trends

Christopher H. Sterling
Timothy Haight

CHILDREN'S TELEVISION: An Analysis of Programming and Advertising

F. Earle Barcus
with Rachel Wolkin

*FREEDOM OF THE PRESS VS. PUBLIC ACCESS

Benno C. Schmidt, Jr.

*THE FUTURE OF PUBLIC BROADCASTING

edited by
Douglass Cater
Michael J. Nyhan

*THE MEDIA AND THE LAW

edited by
Howard Simons
Joseph A. Califano, Jr.

POLITICS IN PUBLIC SERVICE ADVERTISING ON TELEVISION

David L. Paletz
Roberta E. Pearson
Donald L. Willis

*Also available as a PSS Student Edition.